THE ECOLOGICAL CITY

THE ECOLOGICAL CITY

Preserving and Restoring Urban Biodiversity

Edited by Rutherford H. Platt, Rowan A. Rowntree, and Pamela C. Muick

The University of Massachusetts Press   *Amherst*

Copyright © 1994 by

The University of Massachusetts Press

All rights reserved

Printed in the United States of America

LC 93–26506

ISBN 0–87023–883–3 (cloth); 884–1 (pbk.)

Designed by Mary Mendell

Set in Minion by Keystone Typesetting, Inc.

Printed and bound by Thomson-Shore, Inc.

Library of Congress Cataloging-in-Publication Data

The Ecological city : preserving and restoring urban biodiversity /

edited by Rutherford H. Platt, Rowan A. Rowntree, Pamela C. Muick.

p.   cm.

This volume evolved from the 1990 Symposium on Sustainable Cities,

held at the museum of the Chicago Academy of Sciences.

Includes bibliographical references.

ISBN 0–87023–883–3 (cloth : alk. paper) — ISBN 0–87023–884–1 (pbk. : alk. paper)

1. Urban ecology—United States.   2. Urban geography—United States.   3. Urban forestry—

United States.   4. Environmental degradation—United States.   5. Biodiversity—United States.

I. Platt, Rutherford H.   II. Rowntree, Rowan A.   III. Muick, Pamela C.

HT243.U6E26   1994

307.76—dc 20   93–26506 CIP

British Library Cataloguing in Publication data are available.

Drawing on page vii by Modell; © 1990 The New Yorker Magazine, Inc.

This book is dedicated to the memory of

Rutherford Platt 1894–1975

Author, Photographer, Naturalist

"It's so lovely out here you wonder why they have it so far from the city."

Contents

IV   Collaborative Efforts

Conclusion

THE ECOLOGICAL CITY

The Ecological City: Introduction and Overview

Rutherford H. Platt

A World of Cities

A quarter century ago, the demographer Kingsley Davis (1965, p. 41) wrote in *Scientific American:*

> Urbanized societies, in which a majority of the people live crowded together in towns and cities, represent a new and fundamental step in man's social evolution. . . .
>
> Neither the recency nor the speed of this evolutionary development is widely appreciated. Before 1850, no society could be described as predominantly urbanized, and by 1900 only one—Great Britain—could be so regarded. Today, only 65 years later, all industrial nations are highly urbanized, and in the world as a whole, the process of urbanization is accelerating rapidly.

World population has more than tripled during the twentieth century (table 1), thus adding two "worlds" equivalent to the entire global population of 1900 in just nine decades. While it took all of human history until about 1850 to reach one billion, it only took *twelve years* (1975–87) to move from four to five billion people. Even if projected rates of growth decline as expected, the earth's population will double to ten billion by about 2050 (Kates, Turner, and Clark 1990).

A fast-growing proportion of this expanding population now lives in or near cities. Urban population has multiplied *tenfold* in absolute numbers in this century. The proportion of the world's population classified as urban has risen from 14 percent in 1900 to nearly 50 percent (United Nations 1987). This precipitous increase in urban population has resulted from (1) a natural increase in total population (largely due to a declining death rate); (2) the migration of rural population to urban places, and (3) the reclassification of rural settlements to "urban" status or the expansion of the territory of existing cities (Brown and Jacobson 1987, 39).

Table 1 Change in World Population and Urban Population Since 1900

	World Pop.	Urban Pop.	Percent Urban
1900	1.6 billion	224 million	14%
1930	2.0 billion	415 million	20%
1950	2.5 billion	733 million	29%
1970	3.6 billion	1.4 billion	37%
1990	5.1 billion	2.2 billion	43%
2025 (est.)	8.5 billion	5.1 billion	60%

Sources: *World Almanac,* 1991; United Nations, 1987; World Resources Institute, 1990–91, table 5.1

Global urbanization is reflected in a proliferation of new, very large urban regions in Asia, Africa, and Latin America. In 1900 there were only 13 cities exceeding one million, of which all but 3 were in Europe or North America. By 1980, there were about 230 cities worldwide exceeding one million inhabitants, predominantly located in developing countries (World Resources Institute 1986, fig. 3.1B). Seven out of 9 cities exceeding ten million in population in 1980 were in Asia, Latin America, or Africa. In many less developed countries, primate cities have been growing at twice the rate of the national population, in some cases doubling within a decade, as in Mexico City (Teitelbaum 1992/93).

Very large urban regions have grown dramatically in number and absolute population. The thirteen regions exceeding 4 million in 1950 contained 88 million people while the forty-two regions of that size in 1985 held 341 million, of whom 300 million (88%) were in less developed nations (Dogan and Kasarda 1988, 18). For Third World cities, the worst is yet to come: "Of the total increase of nearly 3.8 billion projected over the 1950–2025 period, roughly 2.66 billion (71 percent) will occur after 1990" (World Resources Institute 1990, 66).

The social, economic, and environmental implications of the Third World urban explosion are profound. According to a 1984 United Nations report, characteristics of rapid urbanization include:

—declining environmental quality in urban areas through air, water, and soil pollution, noise, modifications to microclimate, and loss of natural areas;
—severe degradation of the surrounding environment and ecological systems of hinterlands, through urban demand for resources;
—demographic transformations of rural and urban areas through migration with severe social, economic, and environmental consequences;
—inadequate housing, transportation, public services (water, sanitation, schools, health, etc.); resulting in threats to human health and quality of life;

—an urban poor that is vulnerable to deficiencies in food, good water, fuel, and
other basic goods and services;
—the threat of environmental nonsustainability.

These and related threats most immediately afflict squatter or shantytown settlements, which may contain one-quarter to one-half of the total population of
many Third World cities (World Resources Institute 1986, 38). Such districts typically lack basic public services—water, sewerage, paving, and even electrification.
They may occupy sites avoided by earlier development such as tidal flats, unstable
slopes, and ravines subject to flash flooding. This leads to heavy loss of life in the
event of earthquake or hurricane (National Research Council 1987). The inhabitants of such districts chronically suffer from malnutrition, waterborne diseases,
infant mortality, and, in some regions, AIDS.

Toward Managing the Urban Environment

Third World urbanization today is unprecedented in its rate of growth, the absolute numbers of people involved, its geographic scope, and its environmental consequences. However, accelerated urbanization, albeit not at contemporary
rates, also occurred in European and North American cities during the nineteenth
century, as a result of industrialization, international migration, and (later) improvements in public health. Between 1800 and the 1890s, London grew from
860,000 to 5 million, New York from 62,500 to 2.7 million, and Boston from a
provincial seaport of 25,000 to a world city of a half million (table 2).

The compact imperial or colonial urban cores of the late eighteenth century
expanded, physically and legally, into the metropolitan complexes of the twentieth
century. Demands for shelter for factory workers, mill hands, and immigrants led
to the filling of unbuilt areas with ramshackle tenements. In the major cities these
tenement districts were characterized by hideous levels of crowding, inadequate or
nonexistent ventilation, water, and waste removal, and lack of public regulation
(Ashworth 1954, chs. 1–3). These areas were the equivalent of the shantytowns of
the less developed countries today. Meanwhile, the impacts of industrialization in
the nineteenth century, as today, were ravaging the resources of the rural hinterlands, cutting the forests, building over agricultural lands, polluting water resources, and scarring the land with mines and excavations.

Public perception of the conditions and social costs of urban overcrowding was
stimulated by the pioneering research of Sir Edwin Chadwick in England in the
1830s and 1840s. Chadwick used newly available spatial data on illness and infant
mortality to document the relationship between public health and the physical

Table 2   Nineteenth-Century Population Growth of Selected Cities

	1800–1	1850–1	1890–1
London	860,000	1.7 mill.	5.0 mill.
Paris	547,000	1.0 mill.	2.4 mill.
Berlin	201,000	·850,000	1.6 mill.
New York	62,500	660,000	2.7 mill.
Boston	25,000	137,000	448,500

Source: Adapted from Weber (1899/1963)

urban environment. His 1842 "Report of the Poor Law Commissioners Concerning the Sanitary Condition of the Labouring Population of Great Britain" (Flinn 1965) inspired parallel studies by sanitary reformers elsewhere such as the 1848 Griscom report for New York City. A direct result of these investigations was a series of laws, beginning with the British Public Health Act of 1848, that regulated building construction standards and ensured minimum provision of light, air, water, and sewerage to dwelling units. These in turn led in the early twentieth century to the emergence of the modern urban planning movement in Britain, France, and the United States (Peterson 1983), which diffused globally through colonial rule (Platt 1991, 70–76).

While Chadwick and his counterparts focused on the problems of urban squalor, an inquisitive Vermonter, George Perkins Marsh, was investigating human impacts on land and water resources beyond cities. Based on his travels in North America and Europe, and historical research, Marsh's 1864 treatise *Man and Nature, Or Physical Geography as Modified by Human Action*, "was the pioneering attempt to take stock of such transformations and to awaken the public to the magnitude of human impact" (Kates 1987, 529). Marsh's goal as stated in his preface was

> to indicate the character and, approximately, the extent of the changes produced by human action in the physical conditions of the globe we inhabit; to point out the dangers of imprudence and the necessity of caution in all operations which, on a large scale, interfere with the spontaneous arrangements of the organic or the inorganic world; to suggest the possibility and the importance of the *restoration of disturbed harmonies* and the material improvement of waste and exhausted regions. (Marsh 1864/1965, 3, emphasis added)

Drawing on his research in the Mediterranean world, Marsh identified a pervasive pattern of destruction of the "bounties of nature" through ill-considered human intervention in natural processes. Some effects such as deforestation and overgrazing were the direct results of human activities, while others such as accelerated soil erosion were indirect. Marsh reported empirical findings with a moralistic tone that anticipated the environmental rhetoric of recent vintage:

> Man has too long forgotten that the earth was given to him for usufruct alone, not for consumption, still less for profligate waste. . . . Man everywhere is a disturbing agent. Wherever he plants his foot, the harmonies of nature are turned to discords. The proportions and accommodations which insured the stability of existing arrangements are overthrown. Indigenous vegetable and animal species are extirpated, and supplanted by others of foreign origin, spontaneous production is forbidden or restricted and the face of the earth is either laid bare or covered with a new and reluctant growth of vegetable forms, and with alien tribes of animal life. . . . [O]f all organic beings, man alone is to be regarded as essentially a destructive power . . . [against which] nature—that nature whom all material life and all inorganic substance obey—is wholly impotent. (Marsh 1864/1965, 36)

According to Lowenthal (1990, 122), "*Man and Nature* revolutionized environmental thought. It had been conventional wisdom that the earth made man; Marsh showed that man made the earth. Confuting the myth of superabundant and inexhaustible resources, he demonstrated that human impacts were largely unintended, often harmful, and sometimes irreversible."

Marsh's findings reinforced the romantic nostalgia for lost nature reflected in early to mid-nineteenth-century painting (e.g., Cole, Church, Bierstadt), poetry (Longfellow, Whitman, Bryant), and literature (Emerson, Thoreau, Hawthorne) (White and White 1964). These diverse expressions of intellectual reaction to urbanization—scientific and artistic—collectively served as the foundation for the modern conservation movement, or, in Stewart Udall's phrase (1963), "the beginning of wisdom". At the urging of such diverse voices as those of John Muir, John Wesley Powell, and Gifford Pinchot, the United States began to retain and manage portions of its vast Public Domain instead of relinquishing them for private exploitation. The National Forest and National Park systems were established between 1878 and 1916. The designation by New York State of forest reserves in the Adirondacks and the Catskills in the 1890s signified a parallel movement toward the creation of state park systems.

While the immediate results of the preservation movement were thus largely remote from cities, Frederick Law Olmsted, Sr., and his associates brought nature into the heart of many American cities. Beginning with New York's Central Park in 1853, Olmsted demonstrated the value of and need for designed public parks to provide rest and recreation for the crowded urban populace (Sutton 1979). (See further discussion in the next chapter.)

At the turn of the century, the conservation movement and the sanitary reform movement were fused in the paradigm of the "Garden City" as articulated by the British progressive Ebenezer Howard (1902/1965). Howard saw urbanization as an inevitable outgrowth of industrialization and urged that population overspill be accommodated in planned satellite communities affording jobs, good housing, and healthful living conditions, rather than be forced to live in central city slums. Howard's Letchworth Garden City, built in 1902, inspired a number of experimental community projects, and eventually (in much altered form) gave rise to the postwar British New Towns movement (Hall 1982) and equivalent programs in Israel, France, Brazil, Japan, India, Hong Kong, the former USSR, and (less successfully) the United States.

Concurrently with Howard's Garden City movement, Patrick Geddes, a Scottish botanist, was formulating a broader, "ecological" approach to the planning of large cities. In his classic work *Cities in Evolution* (1915), Geddes argued that

> the case for the conservation of Nature and for the increase of our accesses to her, must be stated more seriously and strongly than is customary. Not merely begged for on all grounds of amenity, of recreation, and repose, sound though they are, but insisted upon. On what grounds? In terms of the maintenance and development of life; of the life of youth, of the health of all, . . . and further of that arousal of the mental life in youth, of its maintenance through age, which must be a main aim of higher utilitarianism, and is a prime condition of its continued progress towards enlightenment. (51)

Geddes called for preserving "the remains of hills and moorlands between the rapidly growing cities and conurbations of modern industrial regions" to provide water, recreation, and access to nature for urban residents. He challenged the conventional wisdom regarding the design of city parks: "each with its ring-fence, jealously keeping it apart from a vulgar world. Their layout has as yet too much continued the tradition of the mansion-house drives, to which the people are admitted, on holidays, and by courtesy; and where little girls may sit on the grass" (53).

Anticipating today's interest in fortuitous patches of vacant land, Geddes's "open spaces survey" of Edinburgh disclosed seventy-six small areas totaling ten acres within the "Old Town" that were suitable for gardening or quasi-natural uses.

As with Howard's Garden City, Geddes's theories on urban growth were warmly received by American Progressives, including Lewis Mumford and Benton Mac-Kaye (the "father" of the Appalachian Trail), whose writings would transmit the thinking of Geddes, Howard, and their contemporaries to later generations of urban professionals (Stalley 1972).

American interest in city planning was aroused both by disclosures of continuing squalor in our own cities despite the early sanitary laws (e.g., Riis 1901/1971) and by contemporary planning notions in Great Britain and Germany. The issue of "overcrowding" dominated Progressive proposals of the first decade of the twentieth century, culminating in the seminal First National Conference on City Planning and Congestion in 1909 (U.S. Congress 1910). The American city planning movement and the institution of comprehensive land-use zoning dated from that conference (Toll 1969; Krueckeberg 1983, 69–70).

Despite the "botanical" orientation of Geddes, the development of city planning in advanced nations during the 1920s and 1930s was largely governed by mechanistic considerations of building design, site planning, physical infrastructure, and density. Such functionalism represented a major improvement over the purely aesthetic neoclassicism of the "City Beautiful" designs of the turn of the century, but the relationship of human settlements to the natural world was largely neglected. Only in such utopian constructs as those of Frank Lloyd Wright (Broadacre City) and Le Corbusier (Ville Radieuse) was the need for unbuilt "open space" elevated to a major planning element, albeit in very different forms. But the utopians failed to examine the functions to be performed by open space and to relate its location, extent, and condition to the physical and biological processes of the urban habitat (Platt 1972).

After the hiatus of the Depression and World War II, urbanization resumed in the 1950s, through both the rebuilding of war-damaged cities in Europe and Asia and the construction of new American suburbs fostered by government housing programs and the automobile. The direct effects of urbanization were accompanied by the massive and rapid conversion of rural land through such activities as river basin development, road construction, mineral developments, large-scale commercial agriculture, and deforestation.

The "Man's Role" Symposium

On 16–22 June 1955, seventy-six international scholars convened at Princeton under the auspices of the Wenner-Gren Foundation to examine the implications of "Man's Role in Changing the Face of the Earth." This symposium and its 1,193-page proceedings volume (Thomas 1956) was a scholarly watershed as significant as that defined a century earlier by George Perkins Marsh (to whom the proceedings volume was dedicated). As the 1909 National Conference on City Planning opened the modern era of urban and regional planning, "Man's Role" marked the advent of contemporary research on the impact of human settlements on the biosphere.

While the "Man's Role" symposium addressed an ensemble of problems, regions, and disciplines, a number of its papers stand individually as classic statements. Lewis Mumford and Paul B. Sears in particular addressed broad issues of the ecological sustainability of cities. In "The Natural History of Urbanization," Mumford identified three stages in the evolution of cities in relation to their natural environments: (1) ancient cities such as Babylon that were dependent upon and limited by the resources of their immediate hinterland; (2) imperial cities such as Rome that expanded artificially by preempting food, water, energy and other resources from external sources; and (3) the postindustrial conurbations of Europe and North America. Each successive stage, according to Mumford (1956, 386), has signified "a tendency to loosen the bonds that connect [the city's] inhabitants with nature and to transform, eliminate, or replace its earth-bound aspects, covering the natural site with an artificial environment that enhances the dominance of man and encourages an illusion of complete independence from nature."

The ecological implications of the modern conurbation are dire in Mumford's view:

> Within a century the economy of the Western World has shifted from a rural base, harboring a few big cities and thousands of villages and small towns, to a metropolitan base whose urban spread . . . is fast absorbing the rural hinterland and threatening to wipe out many of the natural elements favorable to life which in earlier stages balanced off against depletions in the urban environment. (Ibid., 395)

Sears (1956) expanded upon this theme of ecological disruption in "The Processes of Environmental Change by Man":

> Least obvious, at any rate to modern urbanized man, is the effect of our present highly complex fauna and flora, organized as they are into communities, upon the environment itself. Through reaction upon habitat these communites, not only insure an orderly cycle of material and energy transformations but also regulate the moisture economy, cushion the earth's surface against violent physiographic change, and make possible the formation of soil. In short, man is dependent upon other organisms both for the immediate means of survival and for maintaining habitat conditions under which survival is possible. (Ibid., 471)

> . . . [T]he changes induced by man whether by sheer destruction or indirectly by accelerating natural processes, are probably more serious to him than the so-called "natural changes" for which he is not responsible. (Ibid., 473)

Away from Managing the Urban Environment

During the thirty-five years since the "Man's Role" symposium, global population has doubled and the number of urban inhabitants has quadrupled (see table 1). Most of the problems recognized in 1955 such as soil erosion, deforestation, water shortages, and natural hazards have worsened in many regions. Meanwhile, new kinds of threats have been identified, such as toxic contamination of groundwater, deforestation, chemical residues in agricultural products, acid rain, photochemical smog, global warming, and depletion of stratospheric ozone (Brown et al. 1989). Since the 1960s, the United States has experienced a series of "movements" addressing urban, environmental, and social problems. Reform efforts have been heavily influenced by the fickle spotlight of the media and the political agenda of the presidential administration in office. These changing tides in turn influence the course of research applied to human problems. Universities have spawned and sometimes abandoned research centers on urban studies, population, environment, energy, global change, and so on.

Paradoxically, as the world becomes more urban, public and scholarly interest in the quality of the human habitat has become increasingly nonurban in focus. In 1965, two contrasting publications marked a watershed of sorts. The first was a special issue of *Scientific American* devoted to "Cities" (from which the Kingsley Davis quote at the beginning of this introduction was taken). The other was the "Future Environments of North America Symposium" and its proceedings volume (Darling and Milton 1966), which virtually ignored cities altogether even though they are the "future environments" of three-quarters of North Americans. Since the mid-1960s, urban settlements have receded as an organizing theme for research programs, conferences, and public policies. Researchers concerned with such topics as aquatic ecosystems, forestry, wildlife management, natural hazards, recreation, and landscape planning have pursued separate and often nonurban agendas. For example, a recent National Academy of Sciences volume on *Biodiversity* (Wilson 1988) devoted only 6 out of 496 pages to urban biodiversity.

The centennial of Clark University in 1987 was commemorated by a major symposium and book entitled *The Earth as Transformed by Human Action: Global and Regional Changes in the Biosphere over the Past 300 Years* (Turner et al. 1990). In the tradition of George Perkins Marsh and the "Man's Role" symposium, the Clark project sought "(1) to document changes [in the biosphere] over the past 300 years; (2) to contrast the global patterns of change to those experienced at the regional level; and (3) to explore the major human forces that have driven these changes" (Turner 1990, xi). The *Earth Transformed* volume is largely concerned with broad changes in the biosphere—land, water, and climate—at various scales, and gives only incidental attention to urbanized regions per se. Ironically, the project thus devoted least attention to the regions that have experienced maximum transformation through human action.

Disinterest in cities per se was also reflected in the United Nations Conference on Environment and Development (the "Environmental Summit") held in Rio de Janeiro in June 1992. According to Alfredo Gastal, director of the United Nations Office of Environment and Human Settlements in Santiago, " 'everyone there was worrying about trees and rain forests, and they were in the city that best exemplifies the worst problem[s] in Latin America, and nothing was said about it' " (*New York Times*, 11 October 1992).

This anecdotal impression is reinforced by an authoritative summary of the Rio Summit in which urban environmental issues are mentioned only indirectly, in terms of externalities inflicted by cities on the biosphere (Haas, Levy, and Parson 1992). Only one out of forty chapters in *Agenda 21*, the official consensus document of the summit, deals with urban issues, under the heading "Sustainable Human Settlements" (United Nations 1992). This chapter appropriately addresses issues of housing, land reform, energy, urban infrastructure and finance, natural hazards, and construction practices. But despite the ecological theme of the Rio Summit, urban ecology is not discussed. The chapter emphasizes technological rather than natural systems within urbanized regions. The precepts of Olmsted, Howard, Geddes, and Mumford regarding the functions of natural systems within cities are forgotten. *Agenda 21* merely exhorts national governments to "develop and support the implementation of improved land-management practices that deal comprehensively with potentially competing land requirements for agriculture, industry, transport, urban development, green spaces, preserves, and other vital needs" (Part I, 75).

Finally, Vice President Al Gore's recent book *Earth in the Balance* also omits any direct discussion of cities and their habitability.

Urban Sustainability

The term "sustainability" has been primarily applied to a variety of nonurban contexts, particularly agriculture, forest management, fisheries, and water resource management (Brown et al. 1987). Much interest in "sustainability" was generated by the *Report of the World Commission on Environment and Development* (Brundtland Commission) (1987), which characterized development as sustainable that "meets the needs of the present without compromising the ability of future generations to meet their own needs." The Worldwatch Institute in its annual *State of the World* reports and the World Resources Institute have contributed to the growing literature on sustainability. Nevertheless, the meaning of the term is both elastic and elusive. It has been termed "an ambiguous concept . . . with no time, space, ecological, technological, or managerial dimension" (O'Riordan 1985).

Sustainability is best understood in an ecological frame of reference. Aldo Leopold in *Sand County Almanac* (1949/1966, pp. 224–25) implied the concept in his famous axiom: "A thing is right when it tends to preserve the integrity, stability, and beauty of the biotic community. It is wrong when it tends otherwise." More elaborately, Brown et al. (1987, 716) define ecological sustainability in terms of "natural biological processes and the continued productivity and functioning of ecosystems. Long-term ecological sustainability requires the protection of genetic resources and the conservation of biological diversity."

Applying an ecological definition of sustainability to urban communities might be viewed as an oxymoron (Greenbie 1990). Urbanization, in the traditional view, destroys natural phenomena and processes, demanding inputs (e.g., food, timber, clean air and water, energy) drawn from elsewhere to replace and augment local resources. The inadequate and impaired "carrying capacity" of the urbanized region is offset by the plundering of nonurban hinterlands, as noted by Mumford in 1956. The ecological impacts of urbanization are experienced far beyond the urban fringe (Platt and Macinko 1983; Cronon 1991). They extend to surrounding agricultural lands, to distant rivers and their watersheds, to lands that provide timber, crops, grazing, water, and recreation, to sources of minerals, to the oceans where wastes are dumped, and to the atmosphere, which is increasingly altered by greenhouse gases and chlorofluorocarbons (CFCs) that emanate from urban sources (Brown and Jacobson 1987, 51).

Urban sustainability thus may be viewed in two senses. The first concerns the protection and restoration of the remaining biological phenomena and processes within the urban community itself—"the greening of the city," in the phrase of Nicholson-Lord (1987). This sense reflects Sears's observation (1956, 471) that ecological communities "insure an orderly cycle of material and energy transformations [and] regulate the moisture economy, cushion the earth's surface against violent physiographic change, and make possible the formation of soil." All but the last of these functions apply to the urban context even more intensely than to the rural. And to these may be added the aesthetic, educational, recreational, and psychological benefits of natural areas within cities. (Olmsted frequently pontificated in his writings on the benefits of large open spaces to the mental and physical well-being of the urban working class.) Edward O. Wilson (1993), although not referring explicitly to cities, recently restated the Olmsted/Sears perspective: "We have only a poor grasp of the ecosystem services by which other organisms cleanse the water, turn soil into a fertile living cover and manufacture the very air we breathe. We sense but do not fully understand what the highly diverse natural world means to our esthetic pleasure and mental well-being."

In the second sense, urban sustainability refers to the impact of cities upon the larger terrestrial, aquatic, and atmospheric resources of the biosphere from which they draw sustenance and upon which they inflict harmful effects. Sustainability in this sense would involve issues of transportation, energy conservation, air and water pollution abatement, material and nutrient recycling, and so forth.

The "Earth Transformed" symposium and the Rio Summit, to the extent that they addressed cities at all, confined their attention to the second category of impacts. Many natural scientists evidently view urban landscapes as essentially doomed and beyond redemption. This book, like the symposium from which it evolved, challenges that conviction. In the tradition of Geddes, it reexamines the functions of and the means of enhancing biotic sustainability within cities. While it is limited geographically to experiences in the United States and Canada, the lessons and insights related should be of broader applicability as the world approaches the 50-percent threshold of urbanization.

The 1990 Symposium on Sustainable Cities

The symposium on "Sustainable Cities: Preserving and Restoring Urban Biodiversity," which led to this volume, was devoted to a reconnaissance of (1) the functions of biodiversity within urban areas, (2) the impacts of urbanization upon biodiversity, and (3) the ways to design cities compatibly with their ecological contexts. The inspiration for the symposium arose initially among members of the Directorate on Urban Ecosystems of the U.S. Man and Biosphere Program (MAB), a unit of the international MAB network organized under the United Nations. The U.S. MAB National Committee was little interested in urban issues, however, and the symposium was independently organized under the direction of the first two editors of this volume and Linda Marquardt of the Chicago Academy of Sciences. Funding was provided by the U.S. Environmental Protection Agency, the U.S. Forest Service, and the National Park Service.

The symposium was held at the museum of The Chicago Academy of Sciences in Chicago's lakefront Lincoln Park. The Academy generously provided staff assistance, conference facilities, and much professional help in the development of the program. Other Chicago area organizations that assisted in the program included the Open Lands Project, the Northeastern Illinois Planning Commission, the Morton Arboretum, the Center for Neighborhood Technology, and the Department of Geography at the University of Illinois at Chicago.

This book, like the symposium that preceded it, is designed to achieve a balance in several respects. First, as befits the subject matter, it is interdisciplinary: geography, ecology, landscape architecture, forestry, wildlife management, and law are represented by the various authors. Second, the book includes articles by scholarly researchers, by private and public program managers, and by citizen activists. Third, articles range from broad generic studies and overview papers (particularly the first three, by Platt, Hough, and Loucks) to reports on specific disciplinary topics and geographic locations. In the latter category are several project case studies on biotic protection and restoration: the DesPlaines River north of Chicago (Hey); the Wildcat and San Pablo creeks in the San Francisco Bay Area (Riley); Portland, Oregon (Poracsky and Houck); Tucson, Arizona (McPherson); the Indiana Dunes (Whitman et al.); the Coachella Valley in Southern California (Beatley); and Lake Tahoe (Goldman).

Most of the papers included in the book were originally presented at the symposium, but those by Beatley, Ahern/Boughton, and Bischoff were subsequently added to the collection. The Riley and Beatley papers are abridged versions of articles that previously appeared in *Environment* and *Environmental Management*, respectively, and are included here with the kind permission of those journals.

The editors regret that the concluding address by William H. Whyte was not available for inclusion in this book because of his illness. He was a stimulating participant throughout the program. Although they are not represented by papers in this volume, the presence at the symposium of "Holly" Whyte, Ann Louise Strong of the University of Pennsylvania, Barrie Greenbie of the University of Massachusetts, and Phil Lewis of the University of Wisconsin at Madison provided a sense of continuity with earlier phases of the urban environmental movement.

Part 1 comprises a trio of overview papers by a geographer, a landscape architect, and a wetlands ecologist, respectively. Rutherford Platt traces the evolution of open space functions and paradigms in American cities, from the colonial commons to the "ecological commons" reflected in contemporary efforts to restore biodiversity within and near urban regions. Michael Hough urges the recognition and protection of indigenous ecological and cultural landscapes in urban design, in place of formalism and "utopian" constructs. Orie Loucks challenges ecologists to transcend mere inventorying of species and to identify the threshold of nonsustainability beyond which ecosystems lose their power to regenerate.

The next two parts address biodiversity issues relating respectively to (1) urban aquatic ecosystems and (2) terrestrial ecosystems, including urban forests, meadows, and arid biomes. There is a considerable literature on the creation and restoration of wetlands in urban areas. The purpose of Part 2 is not to restate what is widely known or available elsewhere (e.g., references cited in the Holland/Prach paper), but to explore selected issues in urban wetlands management as experienced in specific places. Marjorie Holland and Raymond Prach address the question of how ecotones divide wetlands from uplands, in terms of nutrient uptake, assimilative capacity, and considerations of landscape design. Discussion of these and related issues is based on research conducted at a managed wetland site in Regina, Saskatchewan. Donald Hey, an environmental engineer, summarizes his ongoing program to restore a channelized urban creek in the Chicago suburbs to natural condition. Charles Goldman, an internationally respected limnologist, summarizes some findings from his thirty years of research on the impacts of watershed and shoreline development on Lake Tahoe. James Schmid, a consulting ecologist and geographer, raises some practical policy issues relating to the administration of Section 404 of the Clean Water Act (federal wetlands program) with reference to urban wetlands of New Jersey and Pennsylvania.

Part 3 includes several papers on the functions and forms of forests and other upland vegetative communities in urban or urbanizing localities. John Dwyer, Herbert Schroeder, and Paul Gobster report on research at the Morton Arboretum and elsewhere in Chicago regarding the perceptual values attributed to trees and forests by the general public. Gregory McPherson, formerly of the landscape architecture faculty at the University of Arizona and now with the U.S. Forest Service in Chicago, considers the microclimate effects of different landscape types within cities. Jack Ahern and Jestena Boughton address the role of wildflower restoration projects in urban areas and in highway corridors. Richard Whitman and his colleagues, members of the research staff at the Indiana Dunes National Lakeshore, discuss selected examples of research on animal and plant biodiversity in a national park that is severely impacted by human activities. Finally, Annaliese Bischoff reviews evolving public uses and perceptions of a century-old urban park in Springfield, Massachusetts.

Part 4 includes four case studies of intergovernmental and public-private collaboration to promote urban biodiversity and environmental education. Ann Riley, a water planner and community activist, describes the political process that rejected an urban flood-control project planned by the Army Corps of Engineers in favor of a community-designed interagency program to upgrade and restore the riparian habitat in a minority community in the East Bay region of California.

At a larger geographic scale, Timothy Beatley, a planner at the University of Virginia, traces the political process that produced a habitat conservation plan to protect the Coachella Valley lizard from urbanization. Joseph Poracsky and Michael Houck describe the Metro (Portland, OR) multicounty natural resource inventory, which includes computer-based analysis of habitat patches and species diversity identified through remote sensing technology and field investigation. Finally, Karen Hollweg, an environmental educator with the Denver Audubon

Society, summarizes a multicity program funded by the National Science Foundation that focuses upon urban biodiversity wherever it may be found.

References

Ashworth, W. 1954. *The Genesis of Modern British Town Planning.* London: Routledge & Kegan Paul.

Brown, B. J., M. E. Hanson, D. M. Liverman, and R. W. Merideth, Jr. 1987. Global sustainability: Toward definition. *Environmental Management* 11(6): 713–71.

Brown, L., and J. Jacobson. 1987. Assessing the Future of Urbanization. In *State of the world–1987,* ed. L. Brown et al. Washington: Worldwatch Institute.

Brown, L., et al. 1989. *State of the world–1989.* New York: Norton.

Cronon, W. 1991. *Nature's metropolis.* New York: Norton.

Darling, F. F., and J. P. Milton, eds. 1966. *Future environments of North America.* Garden City, NY: Natural History Press.

Davis, K. 1965. The urbanization of the human population. *Scientific American* 213(3): 40–53.

Dogan, M., and J. D. Kasarda. 1988. A world of giant cities. In *The metropolis era,* ed. M. Gogan and J. D. Kasarda. Vol. 1. Newbury Park, CA: Sage Publications.

Flinn, M. W., ed. 1965. *Edwin Chadwick's report on the sanitary condition of the labouring population of Great Britain.* Edinburgh: Edinburgh University Press.

Geddes, P. 1915/1950. *Cities in evolution.* New York: Oxford University Press.

Gore, Al. 1992. *Earth in the balance: Ecology and the human spirit.* Boston: Houghton Mifflin.

Greenbie, B. M. 1990. Synthesizing the oxymoron: The city as a human habitat. Unpublished paper presented at the "Sustainable Cities" symposium, Chicago Academy of Sciences, 4 October 1990.

Haas, P. M., M. A. Levy, and E. A. Parson. 1992. Appraising the earth summit: How should we judge UNCED's success? *Environment* 34(8) (Oct.): 6–11; 26–34.

Hall, P. 1982. *Urban and regional planning.* 2d ed. London: George Allen and Unwin.

Hough, M. 1990. *Out of place: Restoring identity to the regional landscape.* New Haven: Yale University Press.

Howard, E. 1902/1965. *Gardens cities of tomorrow.* Cambridge: MIT Press.

Kates, R. W. 1987. The human environment: The road not taken; the road still beckoning. *Annals of the Association of American Geographers* 77(4): 525–34.

Kates, R. W., B. L. Turner II, and W. C. Clark. 1990. The Great Transformation. In *The earth as transformed by human action,* ed. B. L. Turner II et al.) New York: Cambridge University Press.

Krueckeberg, D. A. 1983. *The American planner: Biographies and recollections.* New York: Methuen.

Leopold, A. 1949/1966. *Sand County Almanac.* New York: Oxford University Press.

Lowenthal, D. 1990. Awareness of human impacts: Changing attitudes and emphases. In *The earth as transformed by human action,* ed. B. L. Turner II et al. New York: Cambridge University Press.

Marsh, G. P. 1864/1965. *Man and nature.* Ed. D. Lowenthal. Cambridge: Harvard University Press.

Mumford, L. 1956. The natural history of urbanization. In *Man's role in changing the face of the earth,* ed. W. L. Thomas, Jr. Chicago: University of Chicago Press.

———. 1961. *The city in history.* New York: Harcourt, Brace and World.

Murphy, D. D. 1988. Challenge to biological diversity in urban areas. In *Biodiversity,* ed. E. O. Wilson. Washington: National Academy Press.

National Research Council. 1987. *Confronting natural disasters: An international decade for natural hazard reduction.* Washington: National Academy Press.

New York Times. 1992. Squalid slums grow as people flood Latin America's cities. October 11.

Nicholson-Lord, D. 1987. *The greening of the cities.* London: Routledge & Kegan Paul.

O'Riordan, T. 1985. Future directions in environmental policy. *Journal of Environment and Planning* A 17:1431–46.

Peterson, J. A. 1983. The impact of sanitary reform upon American urban planning 1840–1890. In *Introduction to planning history in the U.S.,* ed. D. A. Krueckeberg. New Brunswick: Center for Urban Policy Research.

Platt, R. H. 1972. *The open space decision process: spatial allocation of costs and benefits.* Research paper 142. Chicago: University of Chicago Department of Geography.

———. 1991. *Land use control: Geography, law, and public policy.* Englewood Cliffs: Prentice-Hall.

Platt, R. H., and G. Macinko. 1983. *Beyond the urban fringe: Land use issues of nonmetropolitan America.* Minneapolis: University of Minnesota Press.

Riis, J. A. 1901/1971. *How the other half lives.* New York: Dover.

Sears, P. B. 1956. The processes of environmental change by man. In *Man's Role in Changing the Face of the Earth,* ed. W. L. Thomas. Chicago: University of Chicago Press.

Stalley, M. 1972. *Patrick Geddes: Spokesman for man and the environment.* New Brunswick: Rutgers University Press.

Sutton, S. B., ed. 1971. *Civilizing American cities: A selection of Frederic Law Olmsted's writings on city landscape.* Cambridge: MIT Press.

Teitelbaum, M. S. 1992–93. The population threat. *Foreign Affairs* 72(5) (Winter): 63–78.

Thomas, W. L., Jr. 1956. *Man's role in changing the face of the earth.* Chicago: University of Chicago Press.

Toll, S. I. 1969. *Zoned American.* New York: Grossman.

Turner, B. L. II, et al., eds. 1990. *The earth as transformed by human action.* New York: Cambridge University Press.

Udall, S. L. 1963. *The quiet crisis.* New York: Avon Books.

United Nations. 1982. *Estimates and projections of urban, rural and city populations, 1950–2025: The 1980 assessment.* ST/ESA/SER.R/45. New York: United Nations.

———. 1984. *International experts meeting on ecological approaches to urban planning, Suzdal, USSR.: Final report.* Paris: UNESCO/MAB.

———. 1992. *Drafts: Agenda 21, Rio declaration, forest principles.* New York: United Nations.

U.S. Congress. 1910. *City planning*. 61st Cong., 2d Sess. Senate doc. no. 422. Washington: U.S. Government Printing Office.

Weber, A. F. 1899/1963. *The growth of cities in the nineteenth century: A study in statistics*. Ithaca: Cornell University Press.

White, M., and L. White. 1964. *The intellectual versus the city*. Cambridge: Harvard and MIT University Presses.

Wilson, E. O. 1993. "Is Humanity Suicidal?" *New York Times Magazine*. May 30: 24–29.

———, ed. 1988. *Biodiversity*. Washington: National Academy Press.

World Commission on Environment and Development. 1987. *Our common future*. New York: Oxford University Press.

World Resources Institute. 1986. *World resources 1986*. New York: Basic Books.

———. 1990. *World resources 1990–91*. New York and Oxford: Oxford University Press.

I Perspectives on Nature in Cities

From Commons to Commons: Evolving Concepts

of Open Space in North American Cities

Rutherford H. Platt

Introduction

The topic of open space within North American cities has experienced alternating cycles of public attention and neglect. Beginning with New York's Central Park in the 1850s, the urban parks movement inspired by Andrew Jackson Downing and Frederick Law Olmsted, Sr., thrived until the turn of the century. The City Beautiful and Progressive movements fostered a further period of park creation between 1910 and 1940, gaining impetus from New Deal spending and jobs programs during the Depression. In the 1960s, public interest in urban parks and open space was reinvigorated by such influential publications as the report of the Outdoor Recreation Resources Review Commission (1962) (ORRRC) and William H. Whyte's *The Last Landscape* (1968). The ORRRC report led to the creation of the federal Land and Water Conservation Fund in 1965 to support open space acquisition at all levels of government. Federal funding dropped sharply under the Reagan administration, but many states supported open space programs through bond issues. By the end of the 1980s, both federal and state funding had nearly vanished, and many municipalities had become sorely stressed by lack of affordable housing, crumbling infrastructure, solid wastes, air and water pollution, drugs and street violence, and deteriorating tax bases. At the same time, as discussed in the introduction to this book, the focus of the new environmentalism shifted from the city in the 1960s, to the nation in the 1970s, and to the planet in the 1980s, with a consequent weakening of advocacy for traditional urban planning. Nevertheless, several perceptive writers have recently reexamined the social and ecological environments of cities, e.g., Kenneth Jackson in *Crabgrass Frontier* (1985); David Nicholson-Lord in *The Greening of the Cities* (1987); Tony Hiss in *The Experience of Place* (1990); William H. Whyte in *City* (1988); and Michael Hough in *Out of Place: Restoring Identity to the Regional Landscape* (1990). While earlier initiatives focused on public acquisition and management, contemporary thinking on urban open space emphasizes the social and biotic functions of unbuilt or underbuilt land within cities, regardless of ownership.

Unlike British and European cities where large reserves of royal hunting preserves and estates became available for public parks during the nineteenth century, in American cities open spaces have generally been wrested from the private land market through some form of public or private intervention. The objectives sought, the techniques employed, and the forms of open space that resulted have evolved throughout the nation's urban history. Of course, definite chronological divides may not easily be drawn, and individual sites usually serve more than one purpose. While there are many instances of chronological and functional overlap, this article traces the major paradigms and objectives that have motivated retention of urban open spaces over the past two centuries. The story begins and ends with the idea of the commons.

The Colonial Commons

Communal open space in colonial New England such as the Boston Common (fig. 1) emerged from the resolve of early proprietors or inhabitants to retain suitable lands for the town's agrarian and civic purposes. In the case of Boston, the forty-five-acre common was purchased by the town from Reverend William Blaxton in 1634 and thenceforward used for livestock pasture, training grounds for the militia, recreation, burial grounds, and "the hanging of unwelcome Quakers" (Whitehill 1968, 35). The idea of common rights in town lands was derived from the precedent of the English feudal manor, wherein land was allocated in terms of rights of usage based on custom and mutual benefit rather than in terms of freehold ownership (Platt 1991, ch. 3).

Certain New England towns such as Sudbury in the Massachusetts Bay Colony briefly held all of their lands in common, with the inhabitants sharing the use of croplands, pasture, and "waste." During the eighteenth century, however, most lands were allotted to private owners, and town commons were reduced to a few acres of multipurpose open space at the core of each settlement. Hundreds of these commons survive today, but in the form of public lands legally owned and managed by the municipal government rather than held commonly by the inhabitants. Conflict as to appropriate usage of these relict commons arises in many old New England towns, hard-pressed to balance demands for historic preservation, recreation, parking, and municipal revenue. A recent survey by the Connecticut Trust for Historic Preservation of 131 town greens found that 75 were threatened by development of various kinds (*Boston Globe*, 18 October 1992).

The Picturesque Movement and Urban Parks

The Picturesque movement, fostered by landscape architect Andrew Jackson Downing in the 1840s, transplanted another English institution to the United States, namely the deliberately designed "informal" garden. Downing drew inspiration from English landscape gardeners such as Humphry Repton, John Claudius Loudon, and Sir Joseph Paxton (Chadwick 1966, ch. 9; Jackson 1985, ch. 3). The influence of the Picturesque was strongly reflected in cemeteries such as Mount Auburn in Cambridge, Massachusetts, and in private estates such as those designed by Downing in the Hudson River Valley. In his treatise *Landscape Gardening* (1841/1967) and as editor of the influential journal *The Horticulturalist,* Downing called for the establishment of large, lavishly planted urban parks. Unlike the pragmatic spirit of the New England town common, the Picturesque put less emphasis upon functional utility than upon aesthetic effect achieved through landscape design and horticulture. The Boston Public Garden, exemplar of the Picturesque tradition, was created in the 1840s on new land reclaimed from the Charles River estuary bordering the common. Today the public garden, with its eclectic plantings, brilliant flower beds, and swan boats paddling around a tiny pond, offers a romantic, fanciful ambience in contrast to the more prosaic tone of the Boston Common across Charles Street (fig. 2).

Downing and New York newspaper editor-poet William Cullen Bryant in the 1840s called for the creation of a vast "central park" for Manhattan, to be established by municipal action before urban development overwhelmed the entire island. The city acted in time to purchase the necessary land and then held a competition for the design of the park. The winning plan by Frederick Law Olmsted, Sr., and Calvert Vaux generally reflected the influence of Downing and the Picturesque tradition. Their 1858 "Greensward" design artfully blended open meadows, wooded glades, a lake, and varied terrain, largely achieved through extensive earthmoving and planting (fig. 3). Olmsted's goal was not to preserve the somewhat meager natural qualities of the site as it was, but rather to create a pseudorural countryside, "to supply the hundreds of thousands of tired workers, who have no opportunity to spend their summers in the country, a specimen of God's handiwork that shall be to them, inexpensively, what a month or two in the White Mountains or the Adirondacks is, at great cost, to those in easier circumstances" (Olmsted and Kimball 1928/1973, 73).

The simulation of Picturesque rurality dominated most of the subsequent large urban parks designed or influenced by Olmsted (Fabos, Milde, and Weinmayr 1968). A century later, most continue to provide unique oases of open space, despite mounting problems of inadequate maintenance, urban crime, and pro-

Figure 1 The town of
Boston and its Common,
1722

Figure 2   The Boston Public Garden looking toward the Common (photo by the author)

posals for intrusive development. Tony Hiss (1990) acclaimed the illusion of end-less space in the heart of Brooklyn, New York, still provided by the Long Meadow in Olmsted and Vaux's second major project, Prospect Park:

> Remarkably, no buildings are to be seen in this whole expanse; even the horizon, which seems a long way off, beyond the far end of the meadow, shows only trees. Just as remarkably, on almost any day of the year the whole area is flooded with light. The light seems almost to be converging on the meadow from all directions—tumbling onto the grass nearby and also glow-ing through the trees beyond the meadow. (32–33)

Public Health Reforms

Park programs in the second half of the nineteenth century were further motivated by the findings of the nascent public health reform movement stemming from the work of Edwin Chadwick in England. Chadwick's 1842 *Report on the Sanitary Condition of the Labouring Population of Great Britain* (Flinn 1965) inspired coun-terpart investigations into the causes of illness and premature death in the slum districts of American cities (Rosen 1958). While the principal legacy of these re-ports was the adoption of sanitary regulations and building laws, public health reformers also called for the establishment of open spaces and parks to provide the working class with opportunities for fresh air and exercise. Olmsted frequently expressed the view that parks should contribute to the physical and moral well-being of the urban "working man," as mentioned above (Sutton 1971). Tenement districts of large American cities nevertheless grew relentlessly more crowded, with the only available recreation space typically limited to the streets and ventilation courtyards. Urban playgrounds began to appear at the turn of the century, as a result of efforts by Progressive urban reformers such as Jacob Riis and Benjamin Marsh who denounced the evils of urban congestion. Inner city playgrounds, however, apart from the great Olmsted parks, were to remain tragically deficient in number, design, and maintenance.

The City Beautiful

In contrast to both the Picturesque landscape designers and the public health reformers, City Beautiful planners such as Daniel H. Burnham, Charles M. Robinson, and George E. Kessler designed monumental plazas and squares at the hearts of cities to foster civic pride. Beginning with the exuberant "leap backwards" of the 1893 Chicago Columbian Exposition (Giedion 1962), this neoclassical approach to urban design emphasized large, geometric plazas embellished with fountains, statuary, and formal landscaping. The movement transplanted to American cities elements of the grand baroque plans of nineteenth-century imperial European capitals, especially Georges-Eugène Haussmann's redevelopment of Paris and Vienna's Ringstrasse.

The City Beautiful utilized open space as an architectural extension of great public buildings and monuments. From Washington, D.C., to San Francisco and Seattle, the City Beautiful dominated urban downtown architecture and park design from the 1890s until at least the 1950s. The style often seems pompous, impractical, and out of scale with the rest of a city. Peter Hall (1988, 175) has derided City Beautiful monumentalism as applied in "the great commercial cities of middle and western America, where civic leaders built to overcome collective inferiority complexes and boost business."

The 1901 McMillan Commission Plan for the Mall in Washington, D.C., prepared in part by Frederick Law Olmsted, Jr., and Daniel H. Burnham (fig. 4), has endured as a successful legacy of the City Beautiful style. Despite the misgivings of Lewis Mumford (1961, 403–9), who abhorred monumentalism, the Washington, D.C., Mall has functioned well as a "democratic Versailles," perhaps because it is bordered by authentic monuments and national grandeur and is enjoyed by vast numbers of local residents and visitors. Chicago's Grant Park, a legacy of the Burnham and Ben-

Figure 3 Excerpt from the Olmsted and Vaux "Greensward Plan" for Central Park, 1858

nett 1909 *Plan of Chicago*, is another City Beautiful success, due in part to the vitality of Chicago itself and especially its central business district that adjoins the park. On the other hand, the 1903 Burnham design for Cleveland's lakefront mall, an imitation of the Place de la Concorde in Paris, yielded an ensemble of brooding beaux arts government buildings set amid a vast, windy expanse of cement and grass. Monumentality was revived in the 1980s as a motif for corporate headquarters and office parks, combining visual pomposity with ecological sterility (Whyte 1988).

THE MALL
THE M^cMILLAN PLAN 1901

Figure 4 The McMillan Commission Plan for the Mall in Washington, D.C., 1901

The Garden City and Greenbelts

The idea of the Garden City, one of the most influential notions of twentieth-century urban planning, originated in a small tract by the English reformer Ebenezer Howard, first published in 1898 under the title *Tomorrow: A Peaceful Path to Real Reform* and reissued in 1902 as *Garden Cities of To-Morrow*. The Garden City idea was first and foremost a proposal for "the progressive reconstruction of capitalist society into an infinity of cooperative commonwealths" (Hall 1988, 87). Open spaces in several forms were major design elements of the Garden City, as incorporated into the prototypical community of Letchworth, constructed under Howard's guidance at a site thirty miles north of London. Howard and his disciples in England and America eschewed both the artificial, windswept plazas of the City Beautiful and the designed landscapes of Downing and Olmsted in favor of informal, functional kinds of open land more reminiscent of the colonial commons. These spaces included useful, unpretentious community parks and individual garden plots scattered throughout the residential core of the town.

While it inspired such hopeful American experiments as Radburn, New Jersey, and the New Deal Greenbelt Towns program, the Garden City movement more widely influenced the pattern of private suburban development in this country through zoning laws, beginning in the 1920s. Through minimum lot-size requirements, local zoning privatized and enshrined the amenity function of open space, degrading it into an icon of white, upper-middle-class exclusivity.

Another form of open space in the Garden City concept was a greenbelt of quasi-rural land to separate the community from other towns and to ensure residents access to bucolic pastimes and products. Such a greenbelt was established for Letchworth under community ownership and control, although not to the generous size advocated by Howard. Metropolitan greenbelts established after World War II in Great Britain were based on this concept, as applied at a vastly expanded geographic scale.

The United States has no counterpart to the British greenbelts, but the idea of separating urban districts with strands of open spaces may be traced back to Olmsted's prescient 1882 "Emerald Necklace" plan for the Boston park system. The "necklace" consisted of a series of large open spaces connected by parkways and greenways. It was anchored at one end by the Boston Common and Public Garden and at the other by Olmsted's new Franklin Park. The concept of a system of linked natural areas and parks was expanded in Charles Eliot's 1899 plan for the Boston Metropolitan Parks District and the 1929 Massachusetts Bay Circuit Plan prepared by the Trustees of Public Reservations, a private conservation organization.

The 1909 *Plan of Chicago* proposed the establishment of regional forest preserves to maintain urban separation in the suburban areas of Cook County, Illinois (fig. 5). The Cook County Forest Preserve District, created in 1916, has since acquired some seventy thousand acres of forest and grasslands, much of it in narrow strips of stream bottomlands, running north and south across the now heavily developed suburbs of Cook County. Forest preserves have also been established in neighboring counties and in several other U.S. metropolitan areas.

"Spiritual Ecology": The Indiana Dunes

The long struggle to save the Indiana Dunes officially began in 1916 in a challenge by Stephen Mather, the first director of the National Park Service, to preserve the dunes region. The contest between conservationists and industry resulted in the establishment of both the Indiana Dunes National Lakeshore and a federally sponsored industrial harbor in close proximity to each other in 1966. Conflicts between these and other competing uses of the dunes shoreline at the southern tip of Lake Michigan continue to the present time. The long struggle has reflected, according to J. Ronald Engel (1983), a fusion of ecology and spirituality as directed to the preservation of a symbol-laden natural landscape. From the advent of convenient rail access to the dunes in the 1870s, Chicagoans were attracted there for many reasons—artistic, scientific, literary, recreational, and hedonic. Transcending these many interests, according to Engel, was the significance of the Indiana Dunes as a "sacred center" or "hearth," or, in the words of Carl Sandburg, "a signature of time and eternity." The Chicago landscape architect Jens Jensen in the 1920s established two communal shelters in the dunes as prototypical elements of a utopian "Dunes Park": "The radical communitarianism of the Beach House and the Dunes Shelter must be set within the context of the organic dream of human settlement he shared with Louis Sullivan and Frank Lloyd Wright. Jensen pursued this dream in a variety of forms, not least in his designs for Chicago parks. One form intensely absorbing to him was his plan for a permanent settlement in the Dunes" (Engel 1983, 204).

Since the early days of the movement to save Indiana Dunes, grassroots efforts throughout the United States and Canada have attempted, with mixed success, to preserve a diverse array of ecological sites and systems, e.g., redwoods, everglades, prairies, prairie potholes, coastal salt marshes, bottomland hardwoods, old-growth timber, pine barrens. Some of these efforts have been mounted by national and international conservation organizations such as The Nature Conservancy, the Trust for Public Lands, and the World Wildlife Fund, while others are more regional or local in scope. Within urban areas, small patches of woods, swamps, prairie, or even a single tree have inspired passionate protection efforts, but usually in isolation from the larger ecological systems of which these places are a part.

Recreation

As the outward sprawl of American metropolitan regions began in the 1950s and 1960s, the loss of open space in newly developing communities was viewed with

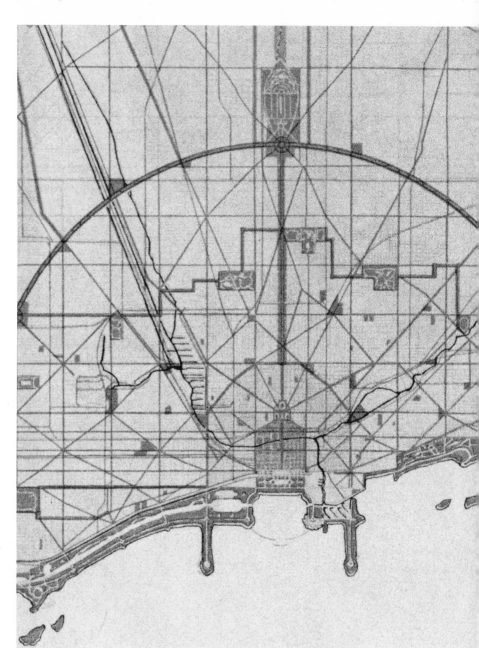

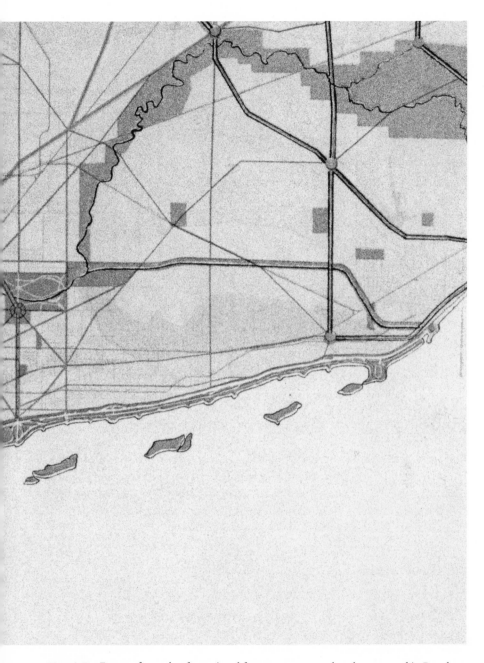

Figure 5   Excerpt from plan for regional forest preserves and parks proposed in Burnham and Bennett's *Plan of Chicago*, 1909

alarm in reports such as *The Race for Open Space* by the Regional Plan Association of New York (1960). The old paradigms of the Picturesque, the City Beautiful, and the Garden City had declined from fashion, and no new concept readily appeared. The report of the Outdoor Recreation Resources Review Commission (ORRRC) in 1962, however, focused new attention on a traditional open-space function, recreation. The ORRRC reported an imbalance between the supply and demand for recreational open space, particularly in quasi-natural settings. The popularity of swimming, fishing, skiing, hiking, and camping and other nonathletic activities suggested the need to conserve public open spaces in and beyond the new suburbs. Indoor sports facilities and parks in central cities were of lower priority than new outdoor recreation and conservation sites, which predominantly served the white middle class (as the photographs in the ORRRC report clearly indicated). The ORRRC report prompted the establishment of a Bureau of Outdoor Recreation in the U.S. Department of the Interior, and the adoption in 1965 of the Land and Water Conservation Fund Act. The fund has since provided over $3 billion to states and local governments on a cost-matching basis to purchase and design open space facilities. (The fund is now administered by the National Park Service.)

Recreational open space within newly developing communities was also facilitated through the refinement of public zoning and subdivision regulation. Respected urban writers such as William H. Whyte, Ann Louise Strong, and Charles Little during the 1960s sought to promote new techniques for retaining open space at the local scale. These mechanisms included cluster zoning, subdivision exactions, planned unit development, floodplain zoning, wetland restrictions, preferential assessment for agricultural land, and scenic or conservation easements. Later, in the 1970s, the transfer of development rights and growth management came into vogue in certain communities as ways to protect open space (Platt 1991). In downtown business districts and new regional malls and office parks, incentive zoning provisions have encouraged the creation of outdoor public spaces. As these facilities are primarily social rather than natural in character, their usefulness depends upon how humanely they are designed (Whyte 1988).

The New Urban Commons

In the 1980s, open space within urban areas assumed a new value for its influence upon urban climate, hydrology, wildlife, and ecological processes, in short, as a new form of urban common resource. This new significance in part reflected the vast geographical expansion of North American cities into their formerly rural hinterlands, expansion fostered by new highway construction and the prevalence of long-distance commuting by automobile. Outlying regional centers at the intersection of major highways have diverted much new growth from the older central business districts and have promoted further extension of city-type activities into areas only recently converted from farm, forest, or desert. Another factor in metropolitan expansion has been the advent of personal computers, fax machines, and computerized information networks, which reduce the need for spatial propinquity in many kinds of economic activities.

The distinction between *urbs* and *rus* has thus become increasingly blurred. Cities that formerly depended upon their rural hinterlands for their needs—food, water, energy, recreation, waste disposal, clean air—now are encroaching upon those same hinterlands. New York City, for instance, for 150 years has drawn its water supply from pure upland sources far from the city, but must now protect the watersheds of its Catskill and Delaware River Basin reservoirs from development resulting from the sprawl of its own metropolitan region (*New York Times*, 20 December 1992).

Even as the city merges with and transforms its rural hinterland, a reverse process may be identified in the gradual "countrification" of the city. As early as 1961, the geographer Jean Gottmann noted that substantial tracts of rural land persisted within the broad parameters of the northeast urban region from Boston to Washington, D.C., which he named Megalopolis. While many of those areas have by now been urbanized, the late twentieth-century American city is not compactly constructed. The centripetal forces that have caused it to overwhelm its hinterlands have also left certain areas temporarily undisturbed as relict woodland, meadow, farm, or wetland within the metropolitan area. Some of these areas have been the focus of preservation efforts, as discussed earlier, and may now be owned by a public agency, land trust, or similar entity.

There is, moreover, a broadening understanding of open space functions and values within urban regions. While specific sites are still sought for anthropocentric activities such as recreation, others are valued for their ecological and hydrologic contributions to the total urban environment, regardless of ownership and management. Natural processes are not limited to formally "preserved" sites. According to David Nicholson-Lord in his book *The Greening of the Cities* (1987, 82), mundane and ignored patches of unused land may yield ecological surprises: "The picture that emerges is thus one of discovery, of an urban society beginning to look at its immediate surroundings with fresh eyes, seeing new possibilities in old things. A radical change in perception is involved."

Open space in the form of cemeteries, railroad yards and rights-of-way, highway borders and medians, utility corridors, and institutional grounds may, with a little neglect of conventional landscaping, retain vestiges of sustainable natural habitat. Vacant lots, tax delinquent properties, former quarries, and riparian borders of minor streams may comprise or revert to pocket wilderness areas (e.g., Duffner and Wathern, 1988, describing a small patch of revegetated land next to the former Berlin Wall).

The naturalness of urban open spaces was antithetical to earlier landscape paradigms. Downing and Olmsted assumed that nature in the city had to be designed, not simply allowed to happen. The City Beautiful planners reduced nature to rows of ornamental trees and shrubs, and plots of grass framed by hedges and pavement. Even the Garden City proponents envisioned open space as fully utilized for civilized purposes, not free to run wild.

Viewed with the "fresh eyes" noted by Nicholson-Lord, the humblest unmanaged residual spaces may provide benefits to surrounding areas in terms of microclimate, drainage, and biodiversity. Even if merely a refuge for strange insects, weeds, and wildflowers, with perhaps a clump of native trees framed against the sky—willow, sycamore, bur oak, spruce, palmetto, or redwood, depending on the region—a small "signature of time and eternity" is retained. This new urban commons of incidental patches of natural habitat cannot reverse the ecological damage inflicted by urbanization, but it may at least soften the impact of the latter. Also, it may remind the urban dweller of the larger biosphere to which the city belongs.

References

Boston Globe. 1992. Townsfolk gather around the greens. 18 October.

Chadwick, A. J. 1966. *The park and the town.* New York: Praeger.

Downing, A. J. 1841/1967. *Landscape gardening.* New York: Funk & Wagnalls.

Duffner, F., and P. Wathern. 1988. Building an urban wilderness: Berlin's green island. *Environment* 30(2): 12–15; 32–34.

Engel, J. R. 1983. *Sacred sands: The struggle for community in the Indiana Dunes.* Middletown, CT: Wesleyan University Press.

Fabos, J. G., T. Milde, and V. M. Weinmayr. 1968. *F. L. Olmsted, Sr.: Founder of landscape architecture in America.* Amherst: University of Massachusetts Press.

Flinn, M. W., ed. 1965. *Edwin Chadwick's report on the sanitary condition of the labouring population of Great Britain.* Edinburgh: Edinburgh University Press.

Giedion, S. 1962. *Space, time and architecture.* Cambridge: Harvard University Press.

Gottmann, J. 1961. *Megalopolis.* Cambridge: MIT Press.

Hall, P. 1988. *Cities of tomorrow.* London: Basil Blackwell.

Hiss, T. 1990. *The experience of place.* New York: Knopf.

Hough, M. 1990. *Out of place: Restoring identity to the regional landscape.* New Haven: Yale University Press.

Howard, E. 1902/1965. *Garden cities of to-morrow.* Cambridge: MIT Press.

Jackson, K. T. 1985. *Crabgrass frontier: the suburbanization of the United States.* New York: Oxford University Press.

Mumford, L. 1961. *The city in history.* New York: Harcourt, Brace and World.

New York Times. 1992. New York City feels pressure to protect precious watershed. 20 December.

Nicholson-Lord, D. 1987. *The greening of the cities.* London: Routledge & Kegan Paul.

Olmsted, F. L., Jr., and T. Kimball, eds. 1928/1973. *Forty years of landscape architecture: Central Park.* Cambridge: MIT Press.

Outdoor Recreation Resources Review Commission. 1962. *Outdoor recreation for America.* Washington: U.S. Government Printing Office.

Platt, R. H. 1991. *Land use control: Geography, law, and public policy.* Englewood Cliffs: Prentice-Hall.

Regional Plan Association of New York. 1960. *The race for open space.* New York: The Association.

Rosen, G. 1958. *A history of public health.* New York: MD Publications.

Sutton, S. B. 1971. *Civilizing American cities: A selection of Frederick Law Olmsted's writings on city landscapes.* Cambridge: MIT Press.

Whitehill, W. M. 1968. *Boston: A topographical history* 2d ed. Cambridge: Harvard University Press.

Whyte, W. H. 1968. *The last landscape.* Garden City, NY: Doubleday.

———. 1988. *City: Rediscovering the center.* New York: Doubleday.

Design with City Nature: An Overview
of Some Issues

Michael Hough

I should begin my discussion with a definition of "sustainable cities." From an environmental design perspective, sustainability has its roots in natural process. In an urban context, sustainability life implies that the products and energy systems of urban life should be passed on to the larger environment as benefits rather than as costly liabilities. This means that man-made works should be designed to produce a net gain in environmental quality and in the overall quality of life. It implies both social and ecological diversity. Sustainability involves the notion that natural systems, influenced by and influencing city form, are as much a part of the urban environment as they are of the "unspoiled" landscape beyond city limits. Urban environments that are sustainable are also place-specific; they belong here, but not there; they are rooted in their particular landscape and, consequently, establish regional identity. People, nature, and places are inherently interdependent—they are part of a common set of issues.

The notion of nature in cities, however, has conventionally been perceived as an oxymoron—like airline food or military intelligence, or even landscape architecture. A personal experience I had some years ago illustrates the problem. A colleague and I had been asked by a local school board to conduct a workshop for science teachers on the advantages of environmental education within the city. The discussion focused on the places close to school such as the ravines, where children could be taken on outings and on regenerating waste places where remnant and naturalized habitats could be found. At one point a sceptical and clearly exasperated teacher asked, "What if there *are* no 'natural places' near the school? How then does one teach children about nature?" To which my colleague immediately replied, "Stand them in the middle of the asphalt school yard and ask them why they are alive. That would be a good beginning."

I remember this incident because it highlights a basic problem about the way many people think about natural systems. The teacher's underlying assumption was that nature is an externality, set apart from human affairs, and that it can only be studied in natural, nonurban surroundings. He was locked into the established doctrine of school curricula everywhere that subjects like geography, biology, or English are to be taught as discrete topics—an educational system that sets the stage for increasing specialization later in life. The problem is encapsulated in Patrick Geddes's well-worn comment that "specialization is the art of knowing more and more about less and less", which suggests that the ultimate goal of knowledge is to know absolutely everything about absolutely nothing. For that teacher, learning about nature had nothing to do with the interdependence of life systems—human and nonhuman—or to the essential quality of immediate experience, which is the means by which children understand the world around them, until they are taught otherwise. The incident also highlighted for me some of the paradoxes and contradictions in the way the city and the larger natural landscape are perceived.

While we all recognize the worldwide implications of forest depletion, water pollution, and the extinction of animal species, the notion that cities themselves are inherently linked to the natural systems that govern life on earth has, until very recently, not been perceived as part of the environmental agenda. In fact, the state of the urban landscapes we live in is a testament to the popular perception that natural processes have little significance for or relevance to design process and form. Ian McHarg's philosophy that the biological disciplines form the indispensable basis for planning the landscape has now, at least in theory, been enshrined by the design professions. Yet the aesthetic paradigms on which much urban design rests, and which are supported by horticultural and engineering technologies, remain the overwhelming determinants that shape the urban landscape.

The Fortuitous Urban Landscape

The economic forces, energy systems, climatic environments, and formal landscapes that have shaped the city too often result in environmental sterility and sensory undernourishment. If we want to find sustainable landscapes in cities, it is not to the lawns, fountains, and flower displays of pedigree parks and formal avenues that we must turn, but to the fortuitous landscapes off the beaten track, in the forgotten and waste places of abandoned industrial areas, in the nooks and crannies behind the corner gas stations, or in the down-at-heel residential areas where the maintenance man doesn't venture. It is here that we find naturalized and evolving urban plant communities, flourishing in the flooded places left after rain; plants, emerging through gratings and cracks in the pavement, that speak to the amazing adaptive power of nature; fortuitous meadows that support a marvelous variety of butterflies, animals, and birds; places where poor drainage systems have created small wetlands that help sustain a stable urban hydrologic balance and beneficial microclimate; sewage lagoons that store the wastes of the city and provide rich and diverse habitat for shorebird populations. The German ecologist Herbert Sukopp, in his studies of bombed-out sites in Berlin, has found new associations of plants evolving in such waste places, and surmises that they may be the prevailing ecosystems of the future (Sukopp et al. 1979). Similarly, the humanized landscape, hidden away in back alleys, on rooftops, and in the backyards of many an ethnic neighborhood, can be described as the product of spontaneous vernacular forces, and provides fascinating and persuasive parallels with the re-generating abandoned lot. Everywhere we look seems to confirm the fact that good design, in the conventional sense, *reduces* environmental diversity while poor design enhances it. It is a paradox of our cities and urban regions that the landscapes we ignore are often more interesting and complex, and have a greater sense of place, than the ones we admire as the expressions of civic pride and good urban design.

If diversity is a necessary basis for environmental and social health, it is the formal city landscapes, such as those left to us from the City Beautiful era, which were imposed over an original natural diversity, that are in need of rehabilitation. The city's fortuitous landscapes provide the basis for understanding the true nature of natural process, and to them design must turn for inspiration. I should like to briefly examine some other related environmental concerns that form a compelling frame of reference for creating a new and more rewarding urban landscape.

Utopian Ideals

Attitudes toward the environment expressed in urban planning and design have, with some exceptions, been more concerned with utopian ideals than with natural process as determinants of urban form. Predetermined images of what places ought to be, rather than what they actually are, dictate form. We find northern cities designed as if they were in California and Californian cities operating on the assumption that they belong to some northerly region of high rainfall; in the sunbelt regions this rainfall is simulated through an elaborate systems of aqueducts that bring water from where it is plentiful to where it is scarce. We see the transfer of natural environments from their places of origin to places where they can be sustained only at high environmental cost—the ideal of a Garden of Eden at the expense of nature, and a prescription for environmental catastrophe. The case of Tucson, Arizona, however, is a compelling example of how this process has been reversed (Hough 1990). The changes that have occurred over the last thirty years or more, transforming the city from a landscape alien to its region to one that reflects it, came about through the need to conserve Tucson's only source of water—

underground aquifers. Legislation introduced in the 1960s requires that ground-water withdrawal be balanced by natural replenishment, and is supported by a range of recycling and educational programs. When I visited Tucson in 1959, the city had the look of a Florida landscape dropped by helicopter into Arizona. Its streets, boulevards, and public parks supported emerald green lawns and lush vegetation—a landscape achieved by intense irrigation and totally alien to the ecology of the surrounding desert. I returned to Tucson in 1987 and found the city transformed. Its once-green public spaces and private gardens had been replaced with an indigenous desert plant community. The city, seen against the bare distant mountains, suddenly had an intense sense of belonging to the landscape that surrounded it. A small but most compelling image of the change was the fact that lawn mowers were nowhere to be seen or heard. What makes rational environmental sense, it would seem, is usually achieved only when change is perceived as absolutely necessary to survival.

Health

Urban health has traditionally been examined from a fixed point of view—that of human health and freedom from disease. If we examine the concept from an environmental perspective, however, quite different conclusions may be drawn about what health means. The concern for urban health dates from the nineteenth century and Benjamin Richardson's utopian plan for the healthy city, which he called Hygeia (Cassidy 1962). The dream was of cities with public baths, swimming pools, and gymnasiums. The streets were to be kept spotlessly clean and "debris was to be carried away beneath the surface." In most Western cities that dream has largely been realized. But this utopian vision, while improving human health, has been accompanied by a progressive deterioration of environmental health. Advances in sanitation that were made to combat disease in the nineteenth-century city began with the development of piped water and stormwater systems, which were later followed by sanitary sewers. With these clearly necessary technological innovations came the attitudes that we now recognize as the throwaway mentality and the inability, or unwillingness, to make connections between social and economic benefits and environmental cost. The benefits of sanitation and well-drained streets that keep one's shoes dry in town have been paid for in destroyed rivers and a degraded larger environment. The economies of using waterways as dumps are paid for in public health threats and closed beaches, and in increased costs of water treatment.

But there have been successes. London's Thames River has been revived and now supports aquatic life and the wildfowl that go with it (Harrison 1976). William

Sopper at Penn State University has, over thirty years, created his sewage treatment systems based on woodland and agricultural soils as living filters, and on an approach that considers effluent and nutrients as resources rather than as products of treatment and disposal (Lull and Sopper 1969). Impoundment systems for storing urban runoff and improving water quality are now common in both U.S. and Canadian cities.

A Multidimensional Role for Parks

The notion that leisure activity is the exclusive function of urban parks derives from the nineteenth-century parks movement. Reformers who included John Claudius Loudon in Britain and Andrew Jackson Downing and Frederick Law Olmsted, Sr., in the United States crusaded for public parks by underlining the health and educational values they would have for working people. They saw contact with nature as a source of pleasure and social benefit, and a necessary way of improving moral standards. Since their inception, the political justification for parks has been driven by considerations of leisure and human health, not their contribution to a healthy environment.

Today, a major role of the parks systems lies in protecting the environment and in maintaining the ecological integrity of the land. They are the essential means for shaping the form and character of urban growth so that woodlands, soils, and wetlands remain undisturbed and able to safeguard the headwaters of urban rivers. Recognition of this role is illustrated by the acquisition of valley lands by the Metropolitan Toronto and Region Conservation Authority. Another example is the protection of Gatineau Park in Canada's National Capital Region in Ottawa.

The German city of Stuttgart has retained the hills that surround the city for parkland and agriculture, because it has found that green hillsides greatly reduce air inversions and pollution problems by maintaining the free flow of katabatic winds that ventilate the city. Parks and open space systems contribute to the integrity of water cycles, the maintenance of water quality, and the control of runoff, the latter a requirement of many municipalities all over the United States and Canada (Winnipeg, Manitoba, and Woodlands, Texas, are two well-known examples). They protect flora and fauna, and conserve natural links throughout the city. They also help to create urban wildernesses, as in London and Toronto.

In addition to concern for the environment, other factors are adding new dimensions to the role of urban parks. It is abundantly clear that concepts of leisure have changed dramatically. There are the needs of the elderly, the physically unpaired, and the large numbers of ethnic groups to be considered. The current preoccupation with physical fitness utilizes the entire city, not just the formal parks, for active recreation. Thus, just as parks have expanded their functions far beyond recreation, so recreation has spatially transcended the boundaries of parks. All this has created demands for a much greater variety of environments than the older, grass-and-trees parks, made for the "average citizen," can provide.

Most cities have vast unused land resources in industrial areas, rights-of-way, and waste places, which remain largely outside the parks system. A major challenge for parks authorities in the 1990s and beyond will lie in institutional change, in devising ways of integrating the total open-space resources of the city into the parks system to serve environmental and social uses and to ensure maximum diversity.

Productivity and the Working Urban Landscape

The city as the center for enormous concentrations of nutrient energy is an anomaly, because urban soils remain almost totally sterile and nonproductive. The working urban landscape involves recycling energy and nutrient resources to productive ends, for example, by using organic matter for urban farming and the enrichment of soils. During the Second World War, British cities produced over 10 percent of the country's total food supply. There were piggeries in Hyde Park and agricultural shows in the basement of Selfridge's department store; urban apiary owners had a lot to say about the kind of trees that should be planted in parks; and the inner city acquired a new unity with the countryside (Hough 1984).

Today, there is the need to find other solutions to hunger in low-income communities besides food banks. The community garden concept is important for food production and because it helps people develop pride in helping themselves. Municipalities have often ignored these issues. A study conducted several years ago compared a map of Toronto's neighborhoods that showed the areas of greatest social and economic need with a map of its city-run allotment gardens. Not surprisingly, these rarely coincided. Political efforts to deal with depressed neighborhoods have been left to a variety of volunteer community organizations. In Britain, the Federation of City Farms has helped establish a new kind of park, known as the City Farms, that integrates allotment gardens, animal husbandry, education, and diverse social activities. In U.S. cities similar work is being done by community organizations such as New York's Green Guerrillas and the Trust for Public Land.

Landscape Restoration

Restoring the landscape to a state of ecological health has become one of the most pressing tasks of the design professions, and is certainly as important as the established North American preoccupation with nature preservation. This task is clearly apparent in landscapes that have been sterilized by piecemeal development, and by municipal design that ignores ecological principles. The rehabilitation of rights-of-way, grass wastelands, and many parks by natural regeneration (re-creating meadows, wetlands, and upland forests) is central to the diversification of urban habitats for human and nonhuman populations.

A new plan, initiated by the citizens of Toronto and supported by the city council, to restore the Don, the most ecologically degraded river in the metropolitan region, is an example of growing citizen involvement in environmental restoration at a large scale (Task Force to Bring Back the Don 1991). In Ottawa, I have been conducting a research program in reforestation to determine the fastest and least costly approaches to restoring native woodland on parkway lands that for years have supported only turf. Preliminary findings after five years were that a woodland canopy of over ten meters in height and the beginnings of forest ground flora can be achieved on the leda clays of the region. Another restoration project has involved the re-creation of forest, meadow, and wetland along the boundaries of a major oil refinery; the restored sites also establish links to an existing creek and lakeshore marsh. Restoration efforts were justified recently when muskrat, fox, and deer appeared at the refinery, having come up along the newly established wooded corridor, much to the delight of the surrounding community.

History and Continuity

The propensity to eliminate natural and cultural heritage as the city spreads outward is nowhere more evident than in new development, and is one of the main reasons why so much urban redevelopment is so featureless. In many places there are no longer any historical reference points by which we can understand where we have come from, as we begin building the new. This is true of landscapes that have lost remnant plant communities, traces of old landform, and geological and cultural features, all of which may still survive in protected or forgotten parts of the city in cemeteries, valleys, worked-out quarries, and depressed residential areas.

The same is true of built form, because redevelopment efforts have traditionally focused on the preservation of significant buildings while ignoring the context that maintains their social or environmental relevance. Today's hot issue—waterfront renewal—is a case in point. The redevelopment process that brings the waterfront back to the people is too often accompanied by the destruction of the previous landscape of industry, work, and natural diversity, and its replacement with glass towers and marine aquariums. The tropical fish tanks and the captive killer whales that leap through hoops and kiss the girl in the bikini for our entertainment and tourist dollars have become the prime attractions of many revived waterfronts, but they tell us nothing about the place. They contribute instead to environmental ignorance, and to a lack of context and identity.

Urban Ecosystem Management

An overall strategy for our cities that brings social and environmental objectives together is urgently needed if the concept of the sustainable city is to be realized. Zurich has done it, at least in terms of land management, by integrating logging and wildlife management, agriculture and recreation, into its park system. This city, in fact, pays for half the cost of its parks maintenance program with the revenues it receives from logs cut in the parks (Hough 1984). The park system is, in a very real economic as well as an environmental sense, sustainable. Essential to such an approach is the notion of "design over time," a view of design based on long-term management that focuses on a continuously evolving and changing urban landscape. This is a very different concept from conventional design and maintenance, whose purpose is to ensure that the created landscape stays as close to what the designer intended as possible.

It will take a long while for this kind of sustainable practice to appear in most North American cities, and, in my experience, current moves to sustainable policies are based less on environmental ideals or ethics than on necessity and what is practical from an economic viewpoint. The task of environmental design in its more profound sense, therefore, is to understand the trends and make the most of the opportunities for beneficial change. The designing of healthy cities begins with a recognition of the inherent human and nonhuman processes that make them what they are. Understanding the nature of places is an essential precursor to making purposeful change, and such change is in my view a far more significant act of creativity than the imposition of prepackaged solutions on the land.

Environmental Education

Environmental literacy lies at the heart of the concept of the sustainable city, which brings me back to my science teacher. To summarize, I see three main issues here. First, perceptions of the city as human environment, separate and distinct from nature, have long been a central problem of education. People don't associate nature with the city environment except as domesticated pets and cultivated plants— poodles, tabby cats, skyrocket junipers, rose gardens, and floral clocks. Second, the idea that the study of life systems can be limited to the scientific method and Linnaean classification militates against what the novelist John Fowles has described as "seeing nature whole", and a sense of wonder (1979). Third, understanding one's home environment is a problem because it is so familiar. Yet understanding the ordinary, representative landscape lies at the heart of this notion of "seeing nature whole." I used to think that knowing your own home place was the key to environmental literacy. I have come to realize, with one of those blinding flashes of the obvious that we like to call insight, that it is only the precursor to the emergence of a larger, and essential, global environmental view.

A more sustainable view of the city and the environment generally is emerging. But I am constantly reminded how difficult it is to change mind-sets about how things should be done. People, like evolutionary processes, are inherently conservative. The need, however, to invest in the protection of nature has never been more urgent. It is tied to notions of environmental and social health, to the essential bond of people to nature, and to the biological sustainability of life itself.

References

Cassidy, J. 1962. Hygeia: A mid-Victorian dream of a city of health. *Journal of the History of Medicine and the Allied Sciences.* April 17.

Fowles, J. 1979. Seeing nature whole. *Harpers Magazine* (November).

Harrison, J., and P. Grant. 1976. *The Thames transformation.* Worcester: The Trinity Press.

Hough, M. 1984. *City form and natural process.* New York: Van Nostrand Reinhold Company, Inc.

———. 1990. *Out of place: Restoring identity to the regional landscape.* New Haven: Yale University Press.

Lull, W., and W. E. Sopper. 1969. *Hydrological effects from urbanization of forested watersheds in the N.E.* U.S. Department of Agriculture Forest Service Paper N.E. 146. Washington: U.S. Department of Agriculture.

Sukopp, H., H.-P. Blume, and W. Kunick. 1979. The soil, flora, and vegetation of Berlin's wastelands. In *Nature in cities.* New York: John Wiley.

Task Force to Bring Back the Don. 1991. Bringing back the Don. Toronto: City of Toronto Department of Planning and Development.

Sustainability in Urban Ecosystems: Beyond an Object of Study

Orie L. Loucks

Introduction

Sustainability is one of those concepts or visions that periodically wash over a society like a storm surge. A mild flooding by more modest concepts may have occurred before, but now almost all the terms we use to describe resources and environments for urban systems must be reassessed in the context of long-term sustainability. This is not simply a concept paradigm shift; rather, the breadth of the dialogue taking place on long-term, intergenerational interests in urban resources is unprecedented.

Such is the sea change as we focus on sustainability in cities. It is no longer useful merely to inventory the ecological resources and processes of the urban environment, or list and summarize what has been ecologically reengineered in the urban system (Loucks 1987). What we need now is to understand and consider the consequences over long periods of modifications being made in urban ecosystems and urban landscape processes. We need to think through spatially distributed effects, downwind and downstream of urban areas, in both the near term and the long term. Indeed, a concern for long-term balance focuses our attention on regenerative capacities of renewable urban elements such as forests, wetlands, streams, and gardens, and their linkage to urban-rural systems. In the process, we have to consider the organic qualities of the city as a system, how it changes temporally and spatially, and how it regenerates. We have to consider how the physical resources of the urban environment interact over long periods with human society and its institutions and commerce.

To elaborate this vision of sustainability in the urban system, I review three topics: (1) background based on studies I did on urban ecology some years ago; (2) factors and processes that are now seen as contributing to impoverishment of urban ecosystems; and (3) two outstanding examples of restoration and recovery. I conclude with some thoughts on how sustainability may be influenced, for better or worse, by the nature of our urban stewardship, and by the approaches to infrastructure and management we have adopted for urban systems.

The theme that runs throughout is that urban ecosystems, with their multi-scaled relationships to surrounding air, water, land, biota, and human institutions, are usually complex and poorly understood. Urban use has required infrastructure that has changed the local hydrology and nutrient exchanges dramatically, altering primary production (e.g., plant growth and tree cover) and the locus of secondary production (e.g., butterflies, possums, and birdlife) and decomposition within landfills or incinerators. For the long term, we now need to know what level of productivity urban ecosystems can maintain if the return process, or feedback, is diverted to external endpoints rather than being recycled locally. Related to these concerns for processing organic material are questions about the living elements of the city, its biological diversity, how it is sustained, and how it contributes services in terms of both production and decomposition in urban systems. Is long-term sustainability possible if organic production is dominated in the short term by a few weedy or opportunistic species, and if decomposition is dominated by relatively few species of insects and the blackbird group? What will be the outcome of ecosystem succession in these systems? I don't know the answers, but let me use several examples to illustrate the questions and summarize what we do know.

Urban Ecology 1970

As part of the U.S. contribution to the International Biological Program, several of us at the University of Wisconsin began in 1969 to test hypotheses about how forested residential areas differ from adjacent natural forests within the Lake Wingra basin, a half-urbanized watershed in Madison, Wisconsin. One half of the watershed was in the city and had been built up during the 1920s, and the other half was an area in the University of Wisconsin Aboretum protected since the early 1930s.

The team of investigators identified ten sub-basins within the Lake Wingra watershed, each representing an independent surface-drainage unit (Cullen and Huff 1972). Eight were linked to the lake by the Madison storm-drainage system, while two drained naturally from the aboretum directly into the lake. Within each of the sub-basins, land use categories (e.g., buildings, roads, woods, etc.) were determined from aerial photographs. Although 50 to 70 percent of the urban sub-basin areas were in a "green" land use, including lawns, gardens, woods, parks, and recreational fields, 88 percent of the two protected watersheds were in natural cover, about half in forest and the remainder in marsh or prairie. Impervious surfaces (buildings, roads, driveways, parking lots, and sidewalks) comprised from 12 percent to 37 percent of the urban sub-basin areas, representing a relatively irreversible alteration of the physical system and associated hydrology.

The changes induced in the soils of the urban watershed through the process of urbanization were among the most surprising. In one residential area studied by Jack Huddleston (1972), approximately 80 percent of the soils had been altered physically and would now be in the early stages of weathering processes moving toward a new soil-chemical equilibrium. Front lawns were more severely altered by the placement of fill than were back lawns, and where undisturbed natural soils were found, they were almost always in the backyard. Most of the fill was believed to have been derived from basement excavations, with only a thin layer of topsoil having been placed on the surface. Weathering and buildup of organic matter to create a normal sequence of soil horizons within the new fill material may require up to a thousand years, and has just begun. Soil weathering will be severely limited by the absence of deeply rooting species within the lawn areas, an important consideration because of the need to increase porosity through the formation and decay of deep woody roots. Regeneration of an equilibrium or sustainable system here requires the pursuit of zero maintenance for flows and percolation, and for the balance of nutrient inputs and losses; it also requires the tolerance by plants of extreme events such as drought or erosion.

The characteristic rates of rainfall infiltration for the soils under forest, prairie (in the arboretum), and lawn surfaces were determined by using a "sprinkling infiltrometer," a device to measure the amount of water entering the soil from a measured quantity applied from a sprinkler. Cumulative infiltration rates on these surfaces showed the natural forest soils to have by far the highest infiltrability (13 cm per hour).

Work comparing the biological productivity of urban and natural forests showed a number of interesting contrasts, with implications for long-term change and potential nonsustainability (Lawson et al. 1972). The trees and shrubs of the urban forest area were generally younger (the oldest trees having died prematurely); in addition, they were more diverse in species composition (due to plantings) and had patchier distribution than natural forest vegetation in the arboretum. The investigators also noted extreme variation in the productivity of urban lawns. This variability appeared to be related to differences in the rates of fertilizer and water application from property to property. Although the urban and natural forest areas showed structural dissimilarities, they nevertheless exhibited comparable characteristics in overall primary productivity at a landscape or large urban area scale (table 1). The annual litter-fall (foliage and herbs/grass) in the natural forest contributed about 430 grams of per square meter dry matter to the soil surface, while in the urban area about 580 grams of dry matter of litter was produced per square meter of permeable surface. Much of this material was removed from the urban area as trash and landfilled, representing a potentially

Table 1   Aboveground Primary Productivity ($g \cdot m^{-2} \cdot yr^{-1}$) for the Noe Woods Oak Forest and the Nakoma Urban Area

Noe Woods (Natural)		Nakoma (Urban)	
Bole	282	Bole	305
Branches	73	Branches	87
Foliage	411	Foliage	320
Herbs	18	Grass	258
Shrubs	28	Shrubs	40
Total	812	Total	1010*

*Data calculated per unit area of permeable (unpaved) surface in the urban area. On a total surface area basis, i.e., including building areas and streets, Nakoma productivity is 776 $g \cdot m^{-2} \cdot yr^{-1}$
Source: Lawson et al. 1972

significant export from the site of local production and a process that is dependent, in the long term, on continuing external nutrient inputs. The overall differences in vegetation also had a direct effect on basin water through their impacts on the permeability and evapotranspiration of stored soil water, and on the redistribution of nutrient stocks, as we shall see.

These studies during the 1969–72 period were among the first to document how the distribution of living components and their support systems (such as water and soil) had been changed structurally by overlaying an urban system on a natural landscape. We barely began to consider how the rates of important processes such as evaporation, fertilizer inputs, soil formation, and organic matter removal had changed, and how these changes would affect the system's long-term sustainability should human inputs ever be withdrawn. Now we do think about such things, but we are still uncertain of the outcomes.

The Impoverishment of Regenerative Capacity: The 1970s to the 1980s

Dislocation of Hydrologic Systems through Urban Engineering

Urbanization has long been recognized as a process that alters the water quality of urban and suburban aquatic systems. Generally this change in land use appears to increase hydrologic and nutrient transport into downstream (receiving) aquatic habitats by increasing the amount of impervious area (thereby increasing runoff), and by increasing the concentration of nutrients in the runoff. A study by Vicki Watson and her colleagues (1981) quantified the impact of urbanization on the nutrient budget of Lake Wingra by determining the seasonal nutrient yield and nitrogen and phosphorous concentrations in the various water sources within the basin, before and after urbanization.

The "before" results were obtained by reconstructing the presettlement water and nutrient budgets of the Lake Wingra basin, using the natural vegetation of the 1,200-acre arboretum as a reference area. Watson and colleagues (1981) assume that the major processes governing hydrologic inputs to the lake have not altered, although the magnitude of some will have changed in relation to the others. Consequently, the same hydrologic transport model could be used to reconstruct presettlement runoff as is used in simulations of the modern watershed dynamics. Parameters affecting the rate and amount of runoff have changed, and presettlement values were estimated from those of the present-day natural areas in the arboretum. Modification in the size of the lake was considered as well as the changes in land use that altered hydrologic and nutrient yields. Storm sewering and diversion of some flows near the east end of the lake have even altered the shape and size of the surface drainage basin (Oakes et al. 1975). Municipal and industrial water wells and subsequent pumpage have had a small effect on groundwater inputs to the lake.

As one might expect, the reconstructed presettlement surface runoff shown in table 2 is substantially less than the present runoff (4.5×10^5 cubic meters vs. 9.9×10^5 cubic meters). Given the assumption of the same average rainfall, the increase in runoff from the past to the present is estimated to have decreased the spring flow and groundwater influx by a somewhat smaller amount, as is evident in the reduction of spring flow by 2.0×10^5 cubic meters (table 2). The cessation of flow from numerous small springs into Lake Wingra since settlement (Baumann et al. 1974) seems to substantiate this decrease, although flows in the major springs and in groundwater inflow remain quite large. The change in rainfall input is due simply to the reduction in the size of the lake by dredging and the filling of wetlands along the shore.

In addition, evapotranspiration (ET) may have been greater in presettlement time because a greater area was covered with vegetation. Since 20 percent of the present basin is covered with impervious surfaces, total basin ET may be only 80 percent of what it once was (i.e., presettlement ET could have been 1.25 times present ET). Results show that for those months in which measurable ET occurs (April through October), the greater presettlement ET is sufficient to account for the difference between past and present runoff, and provides a partial explanation for the smaller modern surface-water budget.

The presettlement nutrient concentrations associated with these hydrologic inflows are a little more difficult to reconstruct. The concentration of nitrogen and phosphorous in the surface runoff of the low-density Nakoma drainage (Prentki et al. 1977) was used to estimate the concentration in presettlement runoff. The modern and presettlement budgets for phosphorus appear in table 3. These tables allow us to assess the relative importance of various sources and different seasons

Table 2   Seasonal Hydrologic Budgets for Presettlement and Present Lake Wingra, WI (10^3 m^3)
(Annual volumes and inputs rounded to two significant digits)

Season	Runoff Volumes		Spring & Groundwater Inputs	
	Prestl.(%)[a]	Present(%)[a]	Prestl.(%)[b]	Present(%)[b]
Spring	350(85)	470(48)	1600(36)	1600(37)
Summer	2.4(0.58)	200(20)	840(19)	840(20)
Fall	23(5.8)	180(18)	710(16)	650(15)
Winter	36(8.7)	140(14)	1300(29)	1200(28)
Annual	410(6.3)[e]	990(16)[e]	4500(69)[f]	4300(69)[f]

[a] Percent of annual runoff that occurs in given season.
[b] Percent of annual springs and groundwater input that occurs in given season.
[c] Percent of annual rainfall that occurs in given season.
[d] Percent of annual inputs from all sources that occurs in given season.
[e] Percent of annual inputs from all sources that derives from runoff.
[f] Percent of annual inputs from all sources that derives from springs and groundwater.
[g] Percent of annual inputs from all sources that derives from rainfall on the lake surface.
Source: Watson et al. 1981

for the changes observed in hydrologic and nutrient budgets. In the bottom rows of tables 2 and 3, Watson and colleagues (1981) show that springs and groundwater are the principal origin of water and nitrogen for Lake Wingra (69 percent for water), while surface runoff is the most important source of phosphorus (42 percent historically compared with 71 percent at present). Springtime is the period of the greatest proportion of nutrient inputs, particularly for phosphorous (47 percent now compared to 56 percent historically), less so for nitrogen (43 percent at both times) and still less for water (34 percent now and 36 percent historically).

The impact of urbanization on the hydrologic and nutrient yields from this watershed (and therefore on the receiving lakes) may be better understood by analyzing the changes in budgets from presettlement to present times. The annual budgets show relatively small changes in total water and nitrogen inputs to Lake Wingra, but a substantial increase in phosphorus input. Most of this increase is attributable to the very large increase in runoff water (although deposition of dust on streets contributes to some increase in the concentration of phosphorus). Phosphorus is a natural leachate from leaves and grass in natural and urban systems, but the tenfold increase in runoff in the summer, when nutrients already are available from internal cycling in the aquatic ecosystem, induce a major alteration of nutrient inputs for receiving waters. The resulting summertime enrichment leads to overproduction of organic matter and anoxia in downstream waters as organisms attempt to break down and recycle the excess production. Nonsustainability becomes evident in the destruction, through oxygen deficiency, of many downstream aquatic environments.

Rainfall Inputs		All Sources	
Prestl.(%)[c]	Present(%)[c]	Prestl.(%)[d]	Present(%)[d]
300(18)	170(18)	2200(34)	2200(36)
600(36)	340(36)	1400(22)	1400(22)
460(28)	260(30)	1200(19)	1100(18)
280(18)	160(18)	1600(25)	1500(24)
1600(25)[g]	920(15)[g]	6400	6200

Effects of Air Pollutants on Urban Ecosystems

The problems inherent in understanding long-term change in urban species and resources are well illustrated by the efforts to understand the stresses and effects of air pollution on urban environments. These effects are only now becoming understood, partly because of their complexity, and partly because of the historical emphasis on the responses of human health to pollutants rather than on the responses of ecosystem health. One of many studies of pollutant effects on lichens in urban areas (McCune 1988) illustrates the nature of the changes involved. Although these studies have focused on a relatively short-lived plant group, lichens, the results are probably applicable to horticultural and other long-lived plant species growing in cities.

The measured properties of the lichen community as studied by Bruce McCune included species composition (the community), species richness (the number of species), and, as described below, a "total cover index" (fig. 1). Air quality monitoring sites already existed in and near Indianapolis and provided air pollution measures during the mid-1980s study period as follows: mean ozone, 65 to 77 micrograms per cubic meter, with a gradient from the southwest (clean air side) to the northeast. Peak one-hour high-ozone values reached 229 to 254 micrograms per cubic meter. The mean annual sulfur dioxide (SO_2) was 23 to 40 micrograms per cubic meter (over seven sites), with the twenty-four-hour high value of 138 to 243. Sulfur dioxide peaks and the highest means tended to be located near the city center. Within the area of greater Indianapolis represented by SO_2 monitors, there was a more or less radially symmetric decline in sulfur levels away from the city center.

Table 3   Seasonal Phosphorus Budgets for Presettlement and Present Lake Wingra, WI (kg of P)[a]

| | Runoff Inputs | | Spring & Groundwater Inputs | |
Season	Prestl.(%)	Present(%)	Prestl.(%)	Present(%)
Spring	250(86)	370(52)	50(30)	50(31)
Summer	1.8(0.62)	120(18)	21(13)	21(13)
Fall	15(5.3)	130(18)	34(20)	32(20)
Winter	22(7.8)	86(12)	63(37)	58(36)
Annual	290(42)	710(71)	170(25)	160(16)

[a] Values in parentheses are percentages and are to be interpreted as in table 2. (Annual totals rounded off)
Source: Watson et al. 1981

Lichens were sampled on trunks of one tree species group, ash, to minimize the variations in substrate, the surface on which lichens grow. The choice of a common species improved the potential for finding suitable trees near air-monitoring sites. Also, McCune reports that ash trees apparently support better-developed lichen communities in polluted urban environments than does any other common urban tree in the midwestern United States. The area covered by lichen species (scored in eight cover classes from 0 to 100 percent) was recorded in two quadrats on each tree (north and south sides), each quadrat approximating the surface of a half cylinder. A total lichen cover index on a tree was then calculated as the sum of the cover classes of all lichens species on that tree.

The differences in lichen species distribution and communities on the ash tree substrates were strongly related to sulfur dioxide levels across the Indianapolis urban area (fig. 1), but not to ozone levels. Lichen species richness was negatively correlated with the SO_2 observation, as was total cover and composition, regardless of whether SO_2 peaks or means were used (McCune 1988). Strong correlations were found and mean annual SO_2 "explained" about 74 percent of the variations along the gradient in the number of lichen species (see r^2 in figure 1). Similarly, mean SO_2 explained 58 percent of the variation in total lichen cover.

These studies show a substantial loss of lichen species in this urban area, a result that has been seen in other cities. Gradually, we are recognizing the same result for higher plants, birds, and other groups of biota. Many lichen species fix nitrogen, but equally important, their biomass forms on rock and bare soil surfaces where no other organic matter can grow. As in the watershed-lake example, where altered nutrient inputs contributed to destabilization of the balance between production and decomposition, here alteration of species composition has the potential to destabilize production, decomposition, and nutrient exchanges in local subsystems. These changes threaten the long-term sustainability of biological processes in urban environments as served by any of the pollution-sensitive species. There is

Bulk Precipitation Inputs		All Sources	
Prestl.(%)	Present(%)	Prestl.(%)	Present(%)
84(37)	48(37)	380(56)	470(47)
68(30)	39(30)	90(13)	180(18)
35(15)	20(15)	84(12)	180(18)
49(18)	23(18)	120(18)	170(17)
230(33)	130(13)	680	1000

no evidence available as to whether the effects being observed are spreading to more species and larger areas at current levels of pollution. More important, the lichen example says nothing about the effects of ozone, a serious human health hazard in many cities, and highly damaging to garden crops, shrubs, and trees. Unlike the SO_2 case just described, downwind rural areas are affected by ozone as much as the central cities are (Loucks and Armentano 1983).

Exacerbation of change through Interactions with Drought, Insects, and Disease

Although the direct effects of changing hydrology and nutrient transport, and of air pollutants are important in inducing changes in the regenerative capacity of urban ecosystems, they do not operate as single factors or in isolation from natural stresses. It remained for the recent research on the effects of acid deposition outside urban areas to examine the "multiple factor" hypothesis and ecosystem

Figure 1   Regression of lichen community parameters against mean annual SO_2 at seven air sampling sites in and around Indianapolis. A: Lichen cover index (sum of cover class values for all species). B: Species richness. Slopes are significantly different from zero at $p < 0.05$. (McCune 1988)

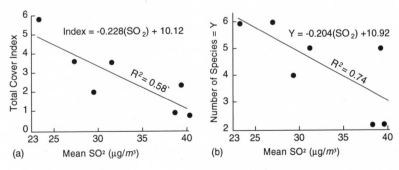

destabilization as explanations for how species and systems respond to subtle long-term stresses. Responses appear to be nonlinear, and sometimes catastrophic.

The phenomenon of a threshold transition to ecosystem destabilization is illustrated by what has happened in the high-altitude spruce forests of Europe (Ulrich 1989). Similar processes are evident from studies of the Ohio Valley oak-hickory forests (Loucks et al. 1991). This research has adopted the perspective of an integrated system rather than of multiple factors, and the hypothesis to be tested involves analysis of "ecosystem destabilization." A systems approach differs from a multiple-factor explanation of "forest decline" in that it focuses on all the major variables exerting control on system dynamics interactively over decades, rather than on "factors" than can be intercorrelated over ten to twenty years. The research examines forest systems with and without air pollution and then asks whether the systems under these two "treatments" respond similarly or differently in the presence of natural stresses brought on by drought, insects, and disease. Another way of considering this question is to ask whether the ecosystem would express a new equilibrium, integrating nutrient cycling rates and tree productivity, after an extended period of natural stresses that occurred *in addition to* the stresses of oxidants and soil acidification. This outcome ultimately must be compared to the response from similar periods of natural stress prior to the introduction of "predisposing" factors related to pollution. The evidence from relatively simple systems, from models, and from U.S. and European field studies is that catastrophic destabilization of the ecosystem is a somewhat frequent outcome. Urban ecosystems experience similar combinations of natural and anthropogenic (human) stress, but analyses have not yet begun to explore questions of local ecosystem destabilization.

In studies using a whole ecosystem approach, knowledge of field processes is used to develop models of nutrient cycling, tree growth, and climate interactions under natural (sustainable) as well as anthropogenically stressed conditions. These models allow the consideration of new expressions for rates of change, time lags, interactions, and the possible development of a new equilibrium among the production and decomposition processes in the ecosystem. The critical hypothesis in considering urban ecosystem sustainability is whether subtle changes in nutrient status and carbohydrate balance (from air pollutants) so alter the tolerance of sensitive species to natural stresses (such as drought, insect outbreaks, and diseases) that the anthropogenic stresses lead to relatively large changes in the system. Such large changes are usually expressed first through a subtle change in the age distribution of biotic populations. All of these conditions and problems exist in urban environments and probably explain some of the losses of arboreal species observed in urban areas today. Reversing these outcomes will be very difficult until the mechanisms are more fully understood.

Restoration of Urban Ecosystems: Steps toward
Regeneration and Sustainability

In spite of the hazards described above, there are many examples of cities where, through determination and ingenuity, urban communities are enhancing and utilizing the immense regenerative capacity of nature. Within limits, it seems possible to reverse much of the spiral of ecosystem impoverishment observed in landscapes supporting high population densities. The turnaround requires technology, however, and a sustained commitment by community leaders, together with a moderate level of capital investment. It means going beyond conventional "set-asides" of open space in the urban area, and going beyond parks as we have known them. Applying the ideas of sustainability to these classical concepts requires that we demonstrate and manage the *regenerative capacity* of the renewable elements in the city. This, ultimately, will mean utilizing the discharges or residuals from the urban system, converting them through recycling to inputs that sustain local subsystems useful to the city as a whole.

Creating a Dragonfly Pond in Yokahama

An example of regeneration in a city is available from Yokahama, Japan. This great industrial city (population 3,200,000) has space for only relatively small parks, so, as in other parts of Japan, landscaping for open space experience has an intensity unique to the culture. This design attribute has been raised to a new level in Honmoku Citizens Public Park in Yokahama, where a simple drainway and elongate concrete detention pond for handling surface water are being converted to a functioning ecosystem, a "dragonfly pond." Japan has unusual diversity in its dragonfly fauna, and these species have special significance for its culture and traditions (Yokahama City 1990).

Restoration of dragonfly breeding habitat was selected as a theme for regenerating the water detention pond for the following reasons: There are several legally protected, uncommon species (such as *Luciola cruciata*) that are now rarely seen, and that might be propagated in a managed habitat. The dragonfly traditionally is viewed as symbolizing the spiritual climate of Japan and therefore presents an intergenerational topic of conversation and admiration. And finally, the habitat of the dragonfly is the emergent and submerged plant life and sediment of a diverse biological system that is potentially compatible with the city. Converting the water detention pond was thought probably to be feasible, and the new pond could enhance other natural amenities of the park.

The dragonfly pond project was implemented as part of an urban parks "refreshing" program being undertaken in Yokahama. It was an attempt to lend personality to some of the otherwise bland urban park facilities. The Honmoku Citizens Public Park was selected for the project partly because the detention pond was readily available, but also because the local community requested restoration of the dragonfly populations. Such a request indicates both the sophistication and receptivity of the people living here and the importance of sustainable living systems to their culture. The public values reflected here are probably related in part to the success of similar initiatives in other parts of the world.

Converting the detention pond to a functioning ecosystem required the creation of a small swamp and stream. The depth of water covering the swamp had to be varied, for part of the season from zero to ten centimeters above baseline, and at other times from zero to twenty centimeters. Water is pumped into the stream at the upper end by a rotary pump that recirculates most of the water from the lowest pool of the pond complex back to the head of the stream. The water then runs through four subponds. Water weeds were planted in the swamp and pond in 1989 according to a specified depth profile for each species. The species chosen were known for their association with the desired type of dragonfly habitat.

Over the first two years of operation, 1988 to 1990, the restoration involved the following steps: sweeping and cleaning the pond, and the first phase of reclaiming and establishing sediment in the "swamp" (1988); the second phase, which was building subponds and planting the aquatic plant species along the stream and in the swamp areas, followed by introducing the nymph stages of the dragonflies (1989); and finally, monitoring the natural changes in the pond's environment through control of water quality, plants, and other aquatic animal life (1990).

One well-known species, *Anax parthenope julius*, and twenty-nine other dragonfly species were stocked in 1989. As of July 1990, fourteen species appeared to have been established. Observing the natural breeding that had begun to take place, the pond managers believed maintenance of thirty species would be feasible. This target is based on a five-year plan. Sustainability, in a narrow sense, is impossible, given the need to circulate the water artificially. But in other ways the system illustrates regenerative capacity; it takes a step toward restoring the balance between production and decomposition, and is being understood and admired as an example of sustainability by thousands of visitors.

The Cuyahoga River Remedial Action Plan at Cleveland, Ohio

Virginia Aveni (1990) describes the Cuyahoga River at Cleveland as a working river. It became an important transportation route in Ohio during the nineteenth century and facilitated massive settlement and industrialization of the area. It still supports the people of the city and the region through transportation, water supply, and other amenities.

Its benefits extend from above Akron, Ohio, to its mouth on Lake Erie. A substantial volume of the headwater flow is diverted and processed into drinking water for the city of Akron, water that is returned to the river downstream via the Akron wastewater treatment plant. Then, for the next thirty-five kilometers (twenty-two miles), the Cuyahoga flows through the Cuyahoga Valley National Recreation Area before entering the heavily industrialized heart of Cleveland. Urban runoff, combined sewer overflows, and other input from within the highly urbanized areas of Akron and Cleveland contribute to the environmental pressures on the relatively small Cuyahoga River. The river was once so contaminated with industrial oil and other wastes that it actually caught fire. However, it now supports a developing recreational area. Night clubs, marinas, and restaurants line both sides of the lower river. Pleasure craft and scullers vie with Great Lakes ore boats for space. As Aveni (1990) says, the fire is out, but much work remains to continue the restoration of the Cuyahoga River.

The Great Lakes Water Quality Board, under the auspices of the U.S./Canada Great Lakes Water Quality Agreement, identified the Cuyahoga River and Cleveland harbor as an "Area of Concern," indicating a binational desire for restoration of water quality. The most impacted area of the river is the shipping channel, where it has been so altered by dredging and shoreline development that virtually no natural riverine habitat remains. Water quality data collected over the last twenty years does indicate improvement, but chemical violations of water quality have been reduced only in the stretch of the river through Cleveland. According to Aveni's description of the two largest sewage treatment plants, the one in Akron and the Cleveland area's Southerly, each may contribute as much as 60 percent of the total flow of the river downstream from their respective discharges.

While the quality of the biological system also has improved in the lower and middle regions of the river, recovery of aquatic organisms has not kept pace with the observed chemical improvements. Populations of benthic invertebrates (insects, snails, worms, and other bottom-dwelling organisms) improved from "poor" to "good" over the past seven years. However, the reestablishment of well-balanced fish communities has not yet occurred. The number of species and total number of fish have increased, but many of the pollution-sensitive species are rare or absent.

Identifying the sources of persistent pollution that remain in the Cuyahoga River is one of the tasks confronting the thirty-five-member Cuyahoga Coordinating Committee (ccc), a planning and advisory body of citizens, government, and industry that oversees preparation of the Cuyahoga River "Remedial Action

Plan." Members of the CCC represent the diverse interests involved, including industry, public interest groups, and federal, state, and local governmental units.

To carry out its task of preparing a plan for remediation of the lower river, the committee has created subcommittees for specific topics, including management of pollution point sources, management of nonpoint sources, biota impairments, toxics consumption by humans, recreational impairments, and socioeconomic considerations. Although the CCC has no powers of enforcement, the subcommittees have been charged with overseeing investigations, preparing status reports on use impairment, analyzing sources and causes of pollution, and generally contributing to the formation of a policy consensus. The committee for community involvement kicked off the creation of the CCC in October 1988 with a boat trip on the river for members of the committee together with Ohio Governor Richard F. Celeste and actor-environmentalist Robert Redford. To involve the public in the planning process, the CCC has established a public charitable organization, the Cuyahoga River Community Planning Organization. This entity will seek financial support from the community to pay for further study of costs and benefits, and to determine how the restoration of beneficial uses can enhance the region's economy and quality of life.

Aveni (1990) reports that although the Cuyahoga Remedial Action Plan is still far from being implemented, the urban community as a whole is looking ahead to the framework for implementing and restoring resilience and repair capability in the urban river ecosystem, key steps to sustainability. As in Yokahama, local support must be in place to provide oversight, and a sense of local ownership is needed for both the plan and resulting restoration to be successful in the long term. The goal is a self-sustaining, renewable ecosystem, characterized by a supportive relationship with the surrounding city in terms of the inputs received as well as the services provided.

Discussion: Urban Stewardship

What do these two examples of successful renewal of urban ecological systems tell us about the infrastructure needed for broad application in management and for urban stewardship generally? For one thing, they tell about the substantial scientific underpinnings required to develop and maintain regenerative capacity in cities. When we intervene to try to restore sustainability, we need to be sure the outcome will be an improvement, not a higher risk of impoverishment. Gaining that assurance requires thorough study of the entire system, as altered and functioning in the urban environment. It also means having suitable measures of what is changing, as well as predictive tools that differentiate between positive and negative outcomes, before we intervene.

The need for quantitative approaches is evident first and foremost in urban wetland management, particularly in the management of locally induced flooding, scouring, and nutrient enrichment of urban wetlands. These topics have been reviewed in a recent paper on the "pulse control" function of wetlands (Loucks 1990), where I show that the pattern of anthropogenically induced flood peaks (and associated sediment loads) in suburban areas can be predicted and mitigated by judicious maintenance of headwater and drier-end wetlands. Thus, sustainability in urban ecosystems may require management of "upstream" environments as well as measures to mitigate downstream deposition of residuals.

In this review of progress toward the stewardship of sustainability in urban ecosystems, I also have recognized two of the apparent sources of impoverishment: hydrologic alterations and air pollution. The former has been associated with excess productivity in downstream environments (beyond the decomposition capacity of these systems), and the latter with declines in the productivity of the nearby terrestrial systems due to direct effects of pollutants on productivity, diversity of species, and availability of biomass. The nutrient enrichment has been viewed, historically, as simply the result of hydrologic alteration, but it now must be viewed as suggesting problems evident in the spatial pattern of human-induced change in an urban landscape. Pollutants released to the air, as well as to the water, add seriously to natural stresses that all plants and animals experience, thereby reducing the potential for renewable, sustainable systems. Both types of effects are correctable through changes in our approach to urban hydrology and air pollution, and some steps have been implemented. At this point, however, they are still insufficient to assure that recovery of the natural resilience and repair capability is underway.

Our community of urban stewards is beginning to understand what is required for successful regeneration of ecosystems: an interested and knowledgeable citizenry, ecosystem technology, public will, political leadership, and appropriate financial support. When all are present, great strides can be taken. When even one or two of these elements are missing, activities in the urban system become compartmentalized and efforts at renewal often fail.

As a conclusion, let me observe that we still need to know much more about the functioning of urban ecological systems, the hierarchical linkages between local and regional processes, and the near-term versus long-term outcomes. Increasing this type of knowledge will require significant new scientific research of a type that has been done in the past only for scattered agricultural systems, for portions of the Great Lakes, and for terrestrial areas such as the Pacific Northwest forests. The new research requires work on the underpinnings of stewardship and policy formulation in ecologically dispersed systems as in urban areas, and on the quantification of the long-term benefits of the management of urban systems. Given how much we have learned about the functioning of urban ecosystems during the past twenty years, it is not unreasonable to predict that we will soon understand more fully what is needed, with sufficient efforts.

References

Aveni, V. 1990. The Cuyahoga: A working river. *Focus on International Joint Commission activities* 15(1): 13–16.

Baumann, P. C., J. F. Kitchell, J. J. Magnuson, and T. B. Kayes. 1974. Lake Wingra 1837–1973: A case history of human impact. *Wisconsin Academy of Science, Arts and Letters* 62:57–94.

Cullen, R. S., and D. D. Huff. 1972. *Determination of land-use categories in the Lake Wingra Basin.* U.S. International Biological Program. Memo report no. 72–43. Institute of Environmental Studies, University of Wisconsin-Madison.

Huddleston, J. H. 1972. *Perturbation of soils in the Manitou Way residential watershed, Lake Wingra Basin, Madison, Wisconsin.* Memo report no. 72–101. Institute for Environmental Studies, University of Wisconsin-Madison.

Lawson, G. J., G. Cottam, and O. L. Loucks. 1972. *Structure and primary productivity of two watersheds in the Lake Wingra Basin.* Memo Report no. 72–98. Institute for Environmental Studies, University of Wisconsin-Madison.

Loucks, O. L., R. T. Prentki, V. J. Watson, B. J. Reynolds, P. R. Weiler, S. M. Bartell, and A. B. D'Alessio. 1977. *Studies of the Lake Wingra watershed: An interim report.* Center for Biotic Systems, Institute for Environmental Studies, University of Wisconsin-Madison.

Loucks, O. L. 1987. The role of basic ecological knowledge in the mitigation of impacts from complex technological systems: Agriculture, transportation, and urban. In *Preserving Ecological Systems,* ed. S. Draggan, J. J. Cohrssen, and R. E. Morrison. New York: Praeger.

———. 1990. Restoration of the pulse control function of wetlands and its relationship to water quality objectives. In *Wetland creation and restoration: The status of the science,* ed. J. A. Kusler and Mary E. Kentula. Washington: Island Press.

———. 1992. Forest response in NAPAP: Potentially successful linkage of policy and science. *Ecological Applications* 2(2): 117–23.

Loucks, O. L., and T. V. Armentano. 1983. Air pollution threats to U.S. National Parks of the Great Lakes region. *Environmental Conservation* 10(4): 303–13.

McCune, B. 1988. Lichen communities along O_2 and SO_3 gradients in Indianapolis. *The Bryologist* 91(3): 223–28.

Oakes, E. L., G. E. Hendrickson, and E. E. Zuehls. 1975. *Hydrology of the Lake Wingra Basin, Dane County, Wisconsin.* U.S. Geological Survey Water Resources Investigations 17–75.

Prentki, R. T., D. S. Rogers, V. J. Watson, P. R. Weiler, and O. L. Loucks. 1977. *Summary tables of Lake Wingra Basin data.* Center for Biotic Systems, Institute for Environmental Studies, University of Wisconsin-Madison.

Ulrich, B. 1989. Forest decline in ecosystem perspective. In *International congress on forest decline research: State of knowledge and perspectives;* ed. B. Ulrich. Karlsruhe, Germany: Forschungsbeirates Waldschaden/Luftverunreinigungen.

Watson, V. J., O. L. Loucks and W. Wojner. 1981. The impact of urbanization on seasonal hydrologic and nutrient budgets of a small North American watershed. *Hydrobiologia* 77: 87–96.

Yokahama City. 1990. Establishment of a dragonfly pond in Honmoku Shimin Public Park. Internal project report.

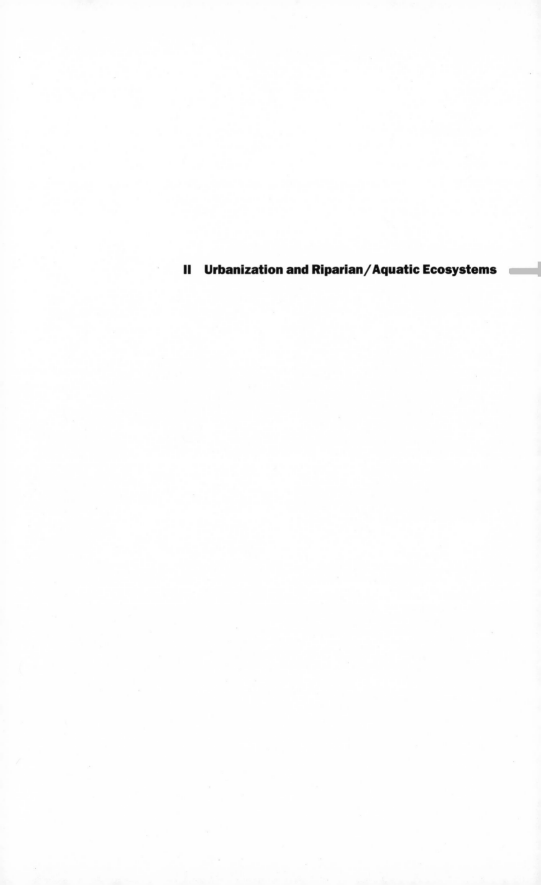

II Urbanization and Riparian/Aquatic Ecosystems

Sustainability of Urban Wetlands

Marjorie M. Holland and Raymond W. Prach

Introduction

This essay addresses the question, What do we need to know to make urban communities environmentally sustainable? In order to answer this question, we must agree on a definition of *sustainable*. We have chosen to interpret *sustainable* to mean that the earth's life-support systems are maintained, and that such maintenance integrates ecological as well as economic uses of these systems. Our focus will be on the sustainability of urban wetlands.

Wetlands are defined as lands transitional between terrestrial and aquatic systems where the water table is usually at or near the surface or the land is covered by shallow water (Cowardin et al. 1979). Wetlands occur over a wide range of hydrologic conditions (Denny 1985; Mitsch and Gosselink 1986; Symoens 1988), and the array of common terms used to describe them has a long history. The classification system most widely accepted in the United States is that developed by Cowardin and colleagues (1979), and that is the system we use here (table 1).

Wetlands perform a wide range of valuable functions for humans (Mitsch and Gosselink 1986; Bedford and Preston 1988). They have local and international significance as regulators of the hydrologic cycle, and they improve water quality (table 2). Wetlands also provide important habitat for freshwater and marine organisms, and they are critical to many bird species as breeding sites and stopovers (staging areas) during migration. It is in part the importance of wetlands to the biological needs of birds that has led to a variety of international agreements addressing wetland conservation and management (Maltby 1986; Hollis et al. 1988).

Wetlands are also important in a landscape context (Forman and Godron 1986; Mitsch and Gosselink 1986). They intercept nutrients and sediment before they reach aquatic ecosystems (Peterjohn and Correll 1984; Correll 1991; Sedell et al. 1991) and provide important habitats for animals, especially those that utilize aquatic and terrestrial ecosystems for migration stopovers, feeding, protection,

Table 1   Wetland Types

Tidal wetland ecosystems	Inland ecosystems
tidal salt marshes	inland freshwater marshes
tidal freshwater marshes	northern peatlands
brackish tidal wetlands	swamps

Source: Mitsch and Gosselink (1986)

and for reproduction. Little is known, however, about the ecotones, or boundaries, between wetlands and other types of ecosystems (Holland et al. 1990).

In inland wetlands, vertical transfers of nutrients, sediment, and energy may occur across at least four surficial boundaries, while horizontal transfers may occur across at least five lateral boundaries (Holland et al. 1990) (see fig. 1). Transfers across lateral boundaries include transfers from the upland to the wetland (upland/wetland ecotone), or from the wetland into open water (wetland/open water ecotone), from groundwater aquifers into soils, or across vegetation zones with each zone dominated by different species (wetland/wetland ecotones). Transfers across surficial boundaries include transfers from aerobic to anaerobic soils, from aerobic soils to surficial vegetation and litter, from vegetation and litter to aerobic soils, from open water to the atmosphere, and from open water to aerobic soils.

All of the aforementioned characteristics also apply to urban wetlands. However, urban wetlands are under pressure from human demands. In general, urban wetlands can be characterized by: (1) restricted water flow, (2) manicured ecotones where margins have been mowed or cleared, (3) increased concentration of human debris and waste products, and (4) absence of mammalian predators. In other words, the life-support systems of urban wetlands have been disturbed or disrupted.

Wascana Centre: Managing an Urban Prairie Wetland

Water and the flora and fauna associated with water are relatively scarce on the prairies. This scarcity led to early recognition of the value of water for human needs, for habitat protection, and for aesthetics. The story of the Wascana Centre in Regina, Saskatchewan, is a small but cogent example of the conflict between management and protection, and the difficulty of coming to grips with the problems inherent in achieving sustainable use of wetlands. This example involves a prairie pothole wetland, representative of a growing number of urban wetlands on the prairies where filling and draining threatens the existence of the entire prairie wetland complex (Turner et al. 1987; Dahl 1990).

Table 2   Wetland Functions and Their Human Utilization (after Hollis et al, 1988)

Role	Elements	Function	Importance to Humankind	Unwise Use
Store/sink	Rare, threatened, or endangered plant and animal species and communities	Genetic diversity, recolonization source	Gene pool, science education, tourism, recreation, heritage	Excessive or uncontrolled harvest, damage, removal, or pollution
	Representative plant and/or animal communities	Ecological diversity, habitat maintenance	Gene pool, science education, tourism, recreation, heritage	Excessive or uncontrolled harvest, damage, removal, or pollution
	Peat	Nutrient, contaminant and energy store, habitat support, water storage	Fuel, Paleo-environmental data, horticultural use, heritage, medical products	Drainage, harvest faster than accumulation, destruction
	Human habitation sites	Archeological remains	Heritage, cultural, scientific, recreation	Destruction, lowering the water table
Pathway	Terrestrial nutrients, water, and detritus	Food chain support, habitat support	Food production, water supply, waste disposal	Interruption or abnormal change of flows, pollution
	Tidal exchanges of water, detritus and nutrients	Food chain support, habitat support, nursery for aquatic organisms	Fish, shellfish, and other food production, waste disposal	Pollution, barriers to flow, dredge and fill
	Animal populations	Support for migratory species, including fish	Harvest, recreation, science	Overexploitation, interruption of migration routes, obstruction, habitat degradation
	Lakes and rivers	Waterways	Navigation	Obstruction, reduced flows and levels

Table 2   *Continued*

Role	Elements	Function	Importance to Humankind	Unwise Use
Buffer	Water bodies, vegetation, soils, and depressions	Flood attenuation	Reduced damage to property and crops	Filling and reduction of storage capacity
	Water bodies, vegetation, soils, and depressions	Detention and retention of nutrients	Food production, improved water quality	Removal of vegetation, drainage and flood protection
	Water bodies, vegetation, soils, and depressions	Groundwater recharge and discharge	Water supply, habitat maintenance, effluent dilution, river fisheries, navigation	Reduction of recharge, overpumping, pollution
	Water bodies and peat	Local and global climate stabilization	Equable climate for agriculture and people	Desiccation
	Water bodies	Large volume, large area	Cooling water	Drainage, filling, thermal pollution
Producer	Production of plants	Food, material, and habitat for migratory species and grazing animals	Harvest of timber, thatch, fuel and food, science, recreation	Overgrazing, overexploitation, drainage, excess change to dry land or agricultural uses
	Animal production	Fish, shellfish, grazing and fur-bearing animals	Harvest and farming	Overexploitation, excess change, habitat degradation
	Organic matter	Methane production, nutrient cycling	Fuel, plant growth	Drainage, desiccation
Sink	Lakes, deltas, floodplains	Sediment deposition and detention	Raised soil fertility, clean downstream channels, improved water quality downstream	Channelization, excess reduction of sediment throughout

Table 2   *Continued*

Role	Elements	Function	Importance to Humankind	Unwise Use
	Lakes, swamps, and marshes	Biochemical self-purification, nutrient accumulation	Natural filter for contaminants, treatment of organic wastes, pathogens and effluents	Destruction of the ecosystem, overloading of the system

Wascana Centre is truly an urban wetland complex, composed of 930 hectares of lawn and natural vegetation including an artificial lake with human-made nesting islands, in the middle of a small but growing city (population 175,000). The lake was created adjacent to a small marsh in 1908 by damming Wascana Creek, a small intermittent creek that flows through the city of Regina. The complex includes a bird sanctuary established shortly after the lake was created, a museum of natural history, the provincial legislative grounds and buildings, and the main campus of the University of Regina (fig. 2).

In the early 1950s, the growth of Regina began to encroach on the land surrounding Wascana Lake, the creek, and contiguous marshes. Fred Bard, then the director of the Saskatchewan Museum of Natural History, took advantage of the community's sense that the water and wetland resource was valuable and needed more vigorous protection. The most effective way to protect a wetland complex—water, marshland, and upland habitat—in Regina during the 1950s was through the Convention for the Protection of Migratory Birds. The convention was signed by the United States and Great Britain (on behalf of Canada) in 1916 in response to extinctions and drastic declines in populations of many bird species. The convention intended the protection of certain "useful or harmless" species including waterfowl and insectivorous birds, but specifically excluded from protection owls, raptors, crows, and other "harmful" species. "Indiscriminate slaughter" was banned, and hunting was regulated. In addition, provisions were made to protect *designated* breeding and staging habitat as bird sanctuaries.

After several years of negotiations among the interested groups, the original bird sanctuary at Wascana Centre, 223 out of 930 hectares (606 out of 2,300 acres), was established as a Federal Migratory Bird Sanctuary in July 1956. The objective in creating a federal bird sanctuary was to add the force of law to the protection and preservation of the bird community in the sanctuary within the wetland complex. In 1962, in recognition of the complex and competing expectations of various groups within the community, the government of Saskatchewan established

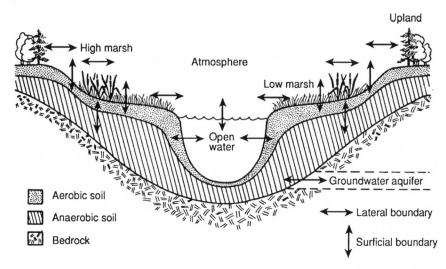

Figure 1 Selected Boundaries of Patches Associated with Generalized Inland Wetlands (after Holland et al. 1990)

through statute the Wascana Centre Authority, an eleven-member committee composed of representatives from the city of Regina, the government of Saskatchewan, and the University of Regina, to manage and develop Wascana Centre.

The wetland complex managed by the authority provides a number of resources. Until recently, the water was used to cool Regina's coal-fired power plant. The Wascana Authority uses the lake and adjacent wetland for its nonpotable needs, primarily irrigation water for lawns, street cleaning, and fire control. However, recreation and tourism are the most visible uses of the area. The two "ponds" of the lake are managed differently, one for bird-watching and environmental education, the other for sailing, rowing, canoeing, skating, cross-country skiing, snowshoeing, and ice boating.

Despite pressure from nearby recreational activities, the wildlife sanctuary retains many of the functions for which it was established. Over three hundred breeding pairs of Canada geese (*Branta canadensis*) nest in and around the sanctuary. Table 3 lists some of the birds that nest in the sanctuary. In spring and fall the complex attracts large numbers of staging waterfowl, particularly Canada geese. More than ten thousand have been recorded in recent counts. Over 115 other species have been observed using the area during migration, including the common loon (*Gavia immer*), white pelican (*Pelecanus erythrorhyncos*), great blue heron (*Ardea herodias*), black-crowned night heron (*Nycticorax nycticorax*), tundra swan (*Olor columbianus*), and Forester's tern (*Sterna forsteri*).

Conflicts sometimes occur between humans and other animals that frequent Wascana Centre. Canada geese and mallard ducks have adapted most readily to life in the complex. Both species began to overwinter in the park when the power plant began operation and created a small area of open water in Wascana Lake that persisted throughout the winter months. Local residents began to feed the birds, which led to more and more birds overwintering. The situation began to resolve itself when the power plant ceased operation in 1978. The overwintering goose population peaked at about 1,500 birds the following winter, then gradually declined to its present level of 300 to 600 birds, as many of the resident birds migrated or died in the natural course of events, and fewer birds overwintered.

However, nesting densities of both mallards and Canada geese remain quite high (table 3). The geese are so prolific that some residents, golfers, and those

Figure 2 Wascana Centre Wetland Complex. 1. Wascana Park; 2. Provincial Legislature; 3. Saskatchewan Centre of the Arts; 4. Douglas Recreational and Fitness Park; 5. University of Regina; 6. Wascana Institute of Applied Arts and Sciences.

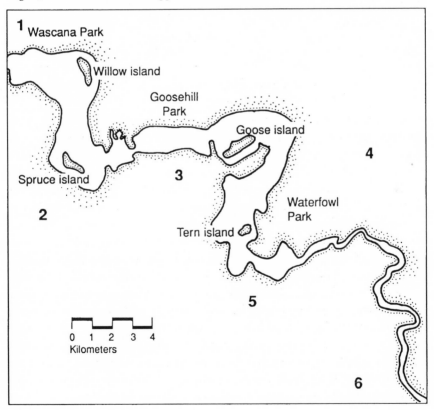

Table 3   Birds Nesting in Wascana Bird Sanctuary

horned grebe *Podiceps auritus*	*
eared grebe *P. caspicus*	100
pied-billed grebe *Podilymbus podiceps*	*
Canada goose *Branta canadensis*	300
mallard *Anas platyrhynchos*	*
pintail *A. acuta*	*
blue-winged teal *A. discors*	*
sora *Porzana carolina*	*
killdeer *Charadrius vociferus*	*
spotted sandpiper *Actitus macularia*	*
American avocet *Recurvirostra americana*	*
Wilson's phalarope *Steganopus tricolor*	*
ring-billed gull *Larus delawarensis*	100
common tern *Sterna hirundo*	*
black tern *Chlidonias niger*	25
short-billed marsh wren *Cistothorus platensis*	*
yellow-headed blackbird *Xanthocephalus xanthocephalus*	*
red-winged blackbird *Agelaius phoeniceus*	*
Brewer's blackbird *Euphagus cyanocephalus*	*

*Numbers not recorded
Source: Wascana Authority (Unpublished data)

wishing to use the park for recreation, consider them and their byproducts a nuisance. Efforts to control the population by removing eggs and replacing them with plastic replicates or destroying eggs or nests have been rejected as unacceptable to many in the community. However, the export of young geese to other areas in transplant programs has proved to be a satisfactory alternative. Eggs and newly hatched goslings have been moved to British Columbia and Quebec in Canada, and to Florida, New Mexico, and other states.

Local farmers are also concerned about the increased number of waterfowl congregating in areas near their crops in preparation for migration. Since waterfowl migration occurs at the same time that grain, the main agricultural product in the area, is prepared for harvest by swathing, management and control of crop damage by mallards, pintails, and geese are continuing activities for the Wascana authority. The authority has opted to provide a lure crop (a grain crop in a field where birds may feed unmolested), so that when local farmers use scare cannons and other techniques to protect their crops, the field-feeding birds have somewhere to go besides another farmer's field (Prach and Surrendi 1977).

As in other wetlands subject to agricultural and urban runoff, the lake and wetland have become eutrophic in response to the increased phosphate runoff from the surrounding terrain. Growth of aquatic plants is so vigorous that it impedes the recreational use of the lake. Weed harvesting may become a regular and expensive management activity. Likewise, the lake is slowly filling with sediment, and the authority is looking for ways to finance a dredging operation to maintain the lake and wetland at a "desirable" successional state.

The complex has a lower biodiversity, with fewer mammals and fewer nesting bird species, compared to rural wetlands of similar size in the area. Many of the mammals, as well as some birds, do not tolerate the presence of humans. Other animals are seen as undesirable. Predators such as skunks, foxes, coyotes, and domestic cats and dogs are actively eliminated from the wetland complex when individual animals threaten birds in the sanctuary. When protected from predators, aggressive and territorial species such as Canada geese thrive. How this affects other species that occur in the sanctuary is not well documented, but it seems likely that the diversity and numbers of individuals of other breeding species that occur naturally elsewhere in the vicinity is lower in the sanctuary.

Although the Wascana Authority has developed a master plan, management decisions about the wetland are made from a local rather than regional perspective. Problems such as those described above are dealt with on an ad hoc basis, and solutions are sought within the boundaries of Wascana Centre. However, many of the problems are generated, in whole or in part, elsewhere in the region, and are indicative of more general ecological questions. For example, both the sedimentation problem and the eutrophication problem may result from lack of a sufficient vegetative buffer zone between the wetland and its tributaries, and the uplands, whether they be farmland or manicured green space. Both problems could be effectively mitigated by planting a greenbelt of perennial vegetation, thus restoring functioning ecotones both within the preserve and along its tributary, Wascana Creek (Tubman 1988; Sedell et al. 1991). In short, the wetland complex (the lake, surrounding land, and its tributary) lacks sufficient ecotones to prevent eutrophication and sedimentation from becoming management problems. At present, resolution of these problems may require extraordinary and expensive remedial action.

Some of the questions faced by the authority have yet to be addressed by researchers in sufficient detail to give guidance in solving management problems. For example, what would be the best mix of landscape features to preserve or restore vertebrate biodiversity within the complex and eliminate the need for managers to control waterfowl populations? Recent evidence suggests that selective planting of different grass species may restrict grazing by Canada geese (Conover 1991). Questions of controlling eutrophication and sedimentation and maintaining biodiversity have not yet been addressed satisfactorily.

The future management of Wascana Centre, located in the prairie provinces, could be affected by global climatic change. Should average temperatures increase, it can be anticipated that evaporation will increase, and the lake level at Wascana Centre will drop. Should temperatures decrease, the growing season would decrease, and the lake at Wascana Centre would be frozen for longer periods each year. Either scenario would have consequences for adjacent agricultural land, higher temperatures putting greater stress on irrigation systems, and lower temperatures forcing farmers to shift to crops tolerant of shorter growing seasons. Thus, any major climatic change may have economic as well as ecological consequences. Current global climate models do not give managers a good sense of what to expect over the next several decades. There is a need for more sensitive models.

Research Needs

The following are a series of possible research questions under two general headings: (1) issues related to the sustainability of wetlands, and (2) issues related specifically to policy and management.

1. Issues related to preserving and restoring sustainable wetlands

   a. *During the growing season, what role does vegetation play in nutrient uptake, especially along upland/wetland boundaries?* There is evidence that vegetation can play a significant role in removing nutrients from the substrate and water column (Holland et al. 1990). We do not know, however, how much this characteristic varies among different types of wetlands and how critical the wetland/upland boundary is. The little data that are available suggest that wetland/upland boundaries may indeed be very important landscape features. For example, recent studies in Maryland (Peterjohn and Correll 1984; Correll 1991) document the role of riparian deciduous hardwood forests in trapping much of the high load of suspended soil particulates and particulate nutrients carried into the forest from upland croplands as storm-induced overland flow.

   b. *How does the assimilative capacity of the wetland boundary compare with the assimilative capacity of the wetland that it bounds?* Most investigations have focused on wetlands, not their boundaries. It appears, however, that the wetland boundary is the most important part of the wetland (Holland et al. 1990). Is this conclusion true, and is it true for all types of wetlands?

   c. *How does the assimilative capacity of upland/wetland ecotones vary in different geomorphological settings and between different types of upland/wetland ecosystems?* Ecotones between uplands and aquatic ecosystems appear to be especially significant landscape features.

d. Along definable hydrologic gradients, are wetland boundaries that are topo-graphically higher more important for trapping and/or converting nutrients than wetland boundaries that are lower?

e. How important are wetland-aquatic ecotones, especially in landscapes where a high percentage of the surface water occurs in lakes and ponds (Holland et al. 1990)? In those situations, is the ratio of wetland area to open water more important than the wetland ecotone?

2. Issues related to policy and management

a. At what level of human investment have ecotones been maintained and restored in the past, and is there any evidence of positive benefits that have resulted from those actions?

b. Do wetland boundaries become less efficient in retaining nutrients when the edge-to-area ratio is altered so that there is less ecotone? If the boundaries are the most important parts of wetlands, then there should be some relationship between the nutrient retention capacity of an entire wetland and the size (area) of the boundary.

c. Can the assimilative capacity of wetlands and wetland boundaries be enhanced or maintained through management? This question is obviously very urgent, yet there is very little information that addresses it directly.

d. At what temporal or spatial scale are research results most useful for decision making and management? Wetland research takes a lot of time to conduct, can be very expensive, and usually considers only one wetland at a time. Most of the questions that are asked of wetlands ecologists, however, deal with issues at several scales, and many of them focus on landscape issues. We think that it would be very productive to undertake a research project that covers several levels and considers both wetlands and wetland ecotones. An example would be to consider all wet-lands in a drainage basin. The goal would be to determine which wetlands are most important in terms of intercepting nutrients and sediments, and which wet-land boundaries are most important. A project of this type will also have the po-tential to identify emergent processes as one moves from the level of a single wet-land boundary to a wetland, and then to a series of wetlands in a drainage system.

Strategies for Future Research

As indicated above, research can be very time-consuming and costly, and the end product may be of limited value to decision-makers. Perhaps the most difficult question to answer is, What type of research should be done in an area of science where the information base is so small, yet the need for information is so great. We believe that field research and simulation modeling are both necessary and that the work needs to be coordinated so that the information will be of use to scientists and resource managers alike. Based on our review, we give the following efforts priority:

1. Conduct simultaneous studies of the roles of upland/wetland ecotones in a variety of landscapes.

2. Characterize the relationships of wetland size, hydrologic characteristics, and dimensions of the ecotone to the assimilative capacity of the wetland and wetland boundaries.

3. Use existing management questions to develop a series of experiments that will test our ability to maintain successfully or enhance the functions of wetland ecotones.

4. Identify traditional, low-intensity management techniques that have successfully maintained or enhanced the functions of wetland ecotones in the past.

5. Utilizing existing descriptive and predictive models used by decision-makers, identify parameters of wetlands and wetland boundaries that need to be better understood. We are suggesting that the models that appear to have the most potential be used to direct the research effort.

Toward a Management of Sustainable Wetlands

As has been suggested by Hollis and colleagues (1988), wise use of wetlands clearly requires action on a large scale, based on a consideration of all factors affecting the wetland. Managing the earth's life-support system, especially in the face of escalating demands on this system, is a national and international imperative (Lubchenco et al. 1991). Research that supports this objective must be at the top of any list of priorities. Recently, the Ecological Society of America has unveiled the Sustainable Biosphere Initiative, which identifies three broad areas as urgent environmental problems for which ecological knowledge is necessary:

—global change, including changes in climate and greenhouse gases, and the links between biotic and abiotic systems;
—biodiversity, including attention to patterns of distribution and loss, and the impact of human activity;
—sustainable ecological systems, including the maintenance of critical natural life-support systems, the management of sustainable systems, and the restoration of damaged systems (Lubchenco et al. 1991).

The Sustainable Biosphere Initiative requires basic research at the intellectual frontiers of ecology, as well as applied research focused on policy-oriented problems. The initiative stresses that primary attention must be given to understanding how systems function and how they respond to stress (Lubchenco et al. 1991). Careful synthesis and integration of research results should ultimately produce local, national, and international policies designed to maintain critical natural life-support systems.

The realization that wetlands cannot be managed in isolation from upstream inputs and the flow of benefits downstream as well as off-site has led to the development of wetland legislation and policies at the local and national level in many developed countries (Holland and Balco 1985; Holland and Phelps 1988; Tubman 1988). In addition to mandating the creation of parks and reserves, legislation may require that alteration of wetlands and stream courses tributary to and downstream from the protected site be subject to regulation (Verhoeven et al. 1988). Most wetlands in developing countries retain a wide range of their natural functions (Gaudet 1979). Many rural economies in Africa and in Southeast Asia are dependent on the utilization of these wetlands. Accordingly, mechanisms for sustainable utilization of wetlands and wetland resources need to be developed and promoted.

Acknowledgments

The authors thank Gerald McKeating, Rutherford Platt, Pam Muick and Lorne Scott for reviews of various drafts of the manuscript.

References

Bedford, B. L., and E. M. Preston. 1988. Developing the scientific basis for assessing cumulative effects of wetland loss and degradation on landscape functions: status, perspectives and prospects. *Environmental Management* 12:751–71.

Conover, M. P. 1991. Herbivory by Canada geese: diet selection and effects on lawns. *Ecological Applications* 1(2): 231–36.

Correll, D. L. 1991. Human impact on the functioning of landscape boundaries. In *Ecotones: The role of landscape boundaries in the management and restoration of changing environments*, ed. M. M. Holland, P. G. Risser, and R. J. Naiman. New York: Chapman and Hall.

Cowardin, L. M., V. Carter, F. C. Golet, and E. T. LaRoe. 1979. *Classification of wetlands and deepwater habitats of the United States.* U.S. Fish and Wildlife Service FWS/OBS-79/31. Washington.

Dahl, T. E. 1990. *Wetland losses in the United States 1780's to 1980's.* Washington: U.S. Fish and Wildlife Service.

Denny, P., ed. 1985. *Ecology and management of African wetland vegetation.* Dordrecht, The Netherlands: Dr. W. Junk.

Forman, R. T. T., and M. Godron. 1986. *Landscape ecology.* New York: John Wiley and Sons.

Gaudet, J. J. 1979. Seasonal changes in nutrients in a tropical swamp: North Swamp, Lake Naivasha, Kenya. *Journal of Ecology* 67:953–81.

Holland, M. M. and J. J. Balco. 1985. Management of fresh waters: input of scientific data into policy formulation in the United States. *Verhandlungen, Internationale Vereingung für Theoretische und Angewandte Limnologie* 22:2221–25.

Holland, M. M., and J. Phelps. 1988. Water resource management: Changing perceptions of resource ownership in the United States. *Verhandlungen, Internationale Vereingung für Theoretische und Angewandte Limnologie* 23:1460–64.

Holland, M. M., D. F. Whigham, and B. Gopal. 1990. The characteristics of wetland ecotones. In *The ecology and management of aquatic-terrestrial ecotones,* ed. R. J. Naiman and H. Decamps. London: Parthenon.

Hollis, G. E., M. M. Holland, E. Maltby, and J. S. Larson. 1988. Wise use of wetlands. *Nature and Resources* 24:2–13.

Lubchenco, J., A. M. Olson, L. B. Brubaker, S. R. Carpenter, M. M. Holland, S. P. Hubbell, S. A. Levin, J. A. MacMahon, P. A. Matson, J. M. Melillo, H. A. Mooney, C. H. Peterson, H. R. Pulliam, L. A. Real, P. J. Regal, and P. G. Risser. 1991. The sustainable biosphere initiative: an ecological research agenda. *Ecology* 72(2): 371–412.

Maltby, E. 1986. *Waterlogged wealth.* London: Earthscan Press.

Mitsch, W. J., and J. G. Gosselink. 1986. *Wetlands.* New York: Van Nostrand Reinhold.

Peterjohn, W. T., and D. L. Correll. 1984. Nutrient dynamics in an agricultural watershed: observations on the role of a riparian forest. *Ecology* 65:1466–75.

Prach, R. W. and D. C. Surrendi. 1977. Crop damage in the Prairie Pothole Region: The problem and future considerations. In *Proceedings of the Second International Waterfowl Symposium.* St. Louis: Ducks Unlimited.

Sedell, J., R. Steedman, H. Regier, and S. Gregory. 1991. Restoration of human impacted land-water ecotones. In *Ecotones: The role of landscape boundaries in the management and restoration of changing environments,* ed. M. M. Holland, P. G. Risser, and R. J. Naiman. New York: Chapman and Hall.

Symoens, J. J., ed. 1988. *Vegetation of inland waters.* Dordrecht, The Netherlands: Kluwer Academic Publishers.

Tubman, L. H. 1988. New Jersey's freshwater wetlands protection act. *Journal of the Water Pollution Control Federation* 60: 176–79.

Turner, B. C., G. S. Hochbaum, and D. J. Naiman. 1987. Agricultural impacts on wetland habitats on the Canadian Prairies, 1981–1985. In *Transactions of the fifty-second North American wildlife and natural resources conference,* ed. R. C. McCabe. Washington: Wildlife Management Institute.

Verhoeven, J. T. A., W. Koerselman, and B. Beltman. 1988. The vegetation of fens in relation to their hydrology and nutrient dynamics. In: *Vegetation in inland waters,* ed. J. J. Symoens. Dordrecht, The Netherlands: Dr. W. Junk.

The Des Plaines River Wetlands Demonstration Project: Restoring an Urban Wetland

Donald L. Hey

> I wish that I could conjure you
>   The stream as it was then;
> Within the glass of memory
>   I see it all again,
>
> But speech is traitor to my wish—
>   Refuses to portray
> Its beauty as I saw it first
> One charming morn in May.
>
> I love, in contemplation sweet,
>   To bring it back once more,
> To watch its sun-tipped waters kiss
> The blue flags near the shore.
>   —E. O. Gale, "The River"

Introduction

The stream of this poem was the Chicago River. The poet knew it as a "modest river with its verdant banks. . . . [I]t curls away to the south a glittering belt of beauty, while the north necklace is lost in umbrageous timber" (Gale 1902). This stream gave the poet many pleasures in his childhood—hunting, fishing, swimming, and the simple enjoyment of an attractive environment. Then came commerce. The ubiquitous lily pads and rushes and wild ducks were displaced from the river by boat harbors and slips and pollutants, which fouled its water to the point that it was almost too putrid to flow. The river, its supporting wetlands, and the surrounding prairie gave way to development—Chicago.

What happened in Chicago was repeated throughout the Midwest. Prior to the settlement of this country by Europeans, wetlands dominated the ecology of the north central states. During the poet Gale's lifetime, from the mid-1800s to the early 1900s, this feature of the midwestern landscape ceased to be. With modern engineering efficiency, tens of millions of acres of land were drained, the prairies were plowed under, and streams and rivers dredged and channelized for drainage and navigation.

These acts of development destroyed wildlife and natural resources and gave rise to flooding and degraded water quality. Yet despite the environmental losses and the flood damage and pollution costs, the economic incentives weighed so heavily in favor of development that more than 50 percent of the wetlands in the lower forty-eight states were drained (Dahl 1990). In Illinois, more than nine million acres of land were drained (Bell 1981), representing 27 percent of the total land area in the state. Today, only 1 to 5 percent of the state's natural wetlands remain (Dahl 1990; Bell 1981).

The poet's lament is as valid today as it was in 1902. We need those rivers, not only to enrich our lives, but also to solve pressing modern, urban problems. We need their natural features to help control flooding, improve and manage water quality, and provide for a greater diversity and abundance of wildlife, as well as for a more attractive environment in which to live, work, and play. These needs will never be met until we are able to return at least some of our urban streams to their natural physical structure, with the supporting wetlands and biological life.

We are not likely to convert our developed lands back to wetlands, to curtail navigation, or to reroute urban drainage. Our nation's economy would not survive. However, the either/or paradigm—economy or environment—must be discarded if we are ever to reach a reasonable accommodation between environmental quality and economic development. We need to sustain our agricultural and industrial activities and, at the same time, improve water quality, reduce flood losses, and encourage wildlife to propagate along our urban and agricultural streams. This, I believe, can be done. It can be done by selectively re-creating aquatic ecosystems along our urban streams, using the original natural wetland as our model (Hey et al. 1982). Urban America has accepted the disturbed agricultural, industrial environment as a condition of economic prosperity. It need not be. Wetlands can be re-created and streams restored to their natural state in those very environments where economic needs first decimated them. What humans have done, they can now undo.

The processes and benefits of restoring aquatic ecosystems are being demonstrated by Wetlands Research, Inc., a Chicago-based, nonprofit organization that manages the Des Plaines River Wetlands Demonstration Project. Before describing the project, however, I would like to tell you about Wetlands Research and, in more detail, the problems that we propose to solve.

Organization and Financing

Wetlands Research, Inc., is a joint venture of Openlands Project and the Lake County Forest Preserve District, formed in 1983 to coordinate the Des Plaines

River project. The governing board is composed of seven members: three elected officials appointed by the president of the forest preserve district (the owner of the project site) and four private citizens appointed by the president of Openlands Project, a Chicago-based conservation organization. Wetlands Research, Inc., is licensed by the Lake County Forest Preserve District to engage in research activities on the project site, including clearing, grading, diverting water, constructing wetland habitats, installing instruments, and conducting experiments. After the research program is completed, the forest preserve district has agreed to maintain the created wetlands.

The project is estimated to cost $18.1 million; almost half of the budget has been secured. Construction is the largest cost component, followed by research, then land acquisition. At the outset, costs were allocated to both public and private interests—the logic being that both would benefit from the results of the project. This allocation remains in force today. In terms of meeting their allocation, state and local contributors rank first and second, private industry is third, and the federal government falls fourth.

Raising the money for construction has been the most difficult task. Few environmental research projects involve such extensive modification of the landscape as the Des Plaines River project. This has often confounded foundations and private donors who are more accustomed to supporting advocacy activities than research or construction projects. The argument for the remaining funds, however, is more compelling because of what has been accomplished to date.

The Economic Imperative and Resulting Problems

The modern American economy was made possible by the efficiency of our agricultural industry. This agricultural efficiency, in turn, required our control of the relationship between land and water. From the earliest days of European settlement, land drainage was not just an accepted practice, it was absolutely essential to the highly productive agricultural lands of the north central United States. In their natural state, these lands were not suited for agricultural production. For all or part of the year, they were inundated. In the 1820s a professor from the University of Pennsylvania, W. H. Keating, observed and described the lands east of Fort Wayne, Indiana: "The country is so wet that we scarcely saw an acre of land upon which settlement could be made. We travelled for a couple of miles with our horses wading through water, sometimes to the girth" (Wooten 1955).

The land, through drainage, was transformed to such an extent that today it bears no resemblance to what Keating saw. A more contemporary traveler made this observation: "[One] who notes the well cared for productive fields, the substantial farm buildings, the good roads and splendid school buildings, may not think that drainage made possible many of the developments. . . . There are more miles of public outlet ditches and drains than there are miles of public highways" (Wooten, 1955).

These were the very wetlands that served the Chicago River, the demise of which was described by the poet Gale. The waters that so quietly stood on these lands were forced off and pushed rapidly downstream, causing flooding, and carrying with them the soils they once protected and the ubiquitous wastes of our agricultural and urban society. Today, much of this land is covered by concrete or asphalt. In Chicago, a city of 228 square miles, 45 percent of the land is now covered by impervious surfaces (National Research Council Committee on Restoration of Aquatic Ecosystems 1992).

While the economic advantage of controlling the relationship between land and water was well recognized, the demise of the wetland did not go entirely unnoticed. Early in this century, a hydraulic engineer with the United States Geological Survey, A. H. Horton, tempered his enthusiasm for land drainage by the following observation:

> The purpose of the drainage is to hurry the waters of excessive rain or melting snow from the land in question. But this act produces an artificial effect on our streams. Through the years of their natural formations the slope and capacity of our rivers have been adjusted by nature to accommodate the flow existing when our swamps were still undrained. These swamps acted as reservoirs, and stored the excessive waters for gradual distribution through the following months of little rain, thus maintaining a fairly uniform flow the year round. Under the new conditions, these reservoirs have been removed and great quantities of drainage water are poured into the streams in a short space of time. Overflowing the banks and flooding of the bottom lands result. Extremes of high and low water and frequency of floods, formerly unknown, are thus brought about. (Horton 1914)

Unfortunately, Horton then proceeded to propose the reclamation of the bottomlands, the ones which were being flooded more frequently, through the use of artificial reservoirs and levees. In his thinking, since the uplands had been reclaimed, the next step was to keep the water off the bottomlands, the riverine wetlands.

Horton's observations were valid, but his conclusions were not. Despite the billions of dollars that have been invested over the intervening years in the very strategies that he proposed, flooding is still a major national hazard, as observed by Stanley Changnon (1985): "[F]lood losses continue to grow despite our major national investments to reduce them; and our approaches for controlling and mitigating flooding have not succeeded." Instead of continuing land reclamation, we should have struck a balance between reclamation and preservation, and set aside lands to store floodwaters. These lands would also have provided wildlife habitat and a better environment for humans.

With the more efficient drainage also came the pollution of our native streams, as typified by the Chicago River. The destruction of the vegetative cover exposed the soil and organic detritus to greater velocities, which entrained the materials, flushing them off of the land and into our rivers. The delicate nutrient balance was greatly disturbed. Our streams became turbid and unattractive, not only for humans but also for the wildlife that was so dependent upon the formerly clear, quiescent waters.

The wetlands had been quietly, unfailingly, performing tasks that were neither needed nor noticed by our ancestors. Water once seeped and flowed through masses of wetland-dependent plants: suspended solids settled and were filtered out, pollutants were cleansed from the waters, and nutrients were absorbed to sustain the luxuriant growth described by early observers. The large volume of water from melted snows and heavy rains slowed its downstream journey as the water spread out across the land, losing its energy, dropping its sediment loads, and returning clean to the river's sinuous channel.

Nor did our ancestors, in their eagerness to develop our land, take any notice of the rich diversity of plants that attracted the multitude of creatures that lived and bred there: fish, waterfowl, and mammals. The abundance of food, the presence of clean water, and the variety of protective cover allowed them to live together. In contrast, the more sterile river habitats resulting from our agricultural and urban development are only able to support low diversity, monotonous flora and fauna.

Like flooding, the problems of river pollution remain unsolved today. We have spent more than $200 billion (U.S. Environmental Protection Agency 1984) to improve the chemical quality of our nation's surface waters. But we have not really made much progress in returning these waters to their former state or even to a state in which a reasonable diversity of aquatic life can exist. In Illinois, for example, despite fifteen years of massive investment and hard work, only 10 percent of formerly contaminated stream reaches have been judged to be suitable for native fauna and flora (Illinois Environmental Protection Agency 1984). This was accomplished at a cost of $13 million a mile.

The past investment strategy for pollution control, much like that for flood control, has often been misdirected. Funding for the Clean Water Act is only directed toward point-source pollution. Contaminates that drain off the expanses of asphalt and cropland, termed nonpoint-source pollution, remain uncontrolled. In a highly urbanized area such as northeastern Illinois, nonpoint-source contaminates account for more than half of the pollutants found in the local stream and rivers. We must develop strategies and generate funding for the control of nonpoint-source pollution if we are to achieve the goals of the Clean Water Act, that is, "to restore and maintain the chemical, physical and biological integrity of our nation's waters."

Re-Creating Wetlands

Clearly, in the past, the economic worth of draining wetlands and despoiling our streams and rivers was greater than the alternative—preservation. As stated by the noted resource economist John Krutilla, "Allocation of any resource to one use precludes its simultaneous use for any incompatible alternative uses. The problem is the substance of economic choice. In the case of the wetlands, however, a misallocation is likely" (Goldstein 1971).

This misallocation occurs for a variety of reasons. First, the beneficiary of the drained land, by definition, does not suffer from the resulting floods. Also, in parts of the United States where water is not particularly scarce, the contamination of any one source has not been of great concern. Finally, wildlife resources are, in Krutilla's terms, "fugitive resources," moving from place to place. They were of only limited value to the early settlers, and were easily replaced by substitute products such as domestic animals.

Today, our values are different. We are more aware of the cause-and-effect relationship between our actions and the fragile environment in which we live. We care about the quality of our environment, in measurable terms as well as in terms of aesthetics. The past misallocation of our wetland resources can be redressed, for the marginal value of the re-created environments will be greater than the alternatives—flooding and pollution.

Can wetlands be re-created to imitate their natural prototypes? I believe the answer is yes, at least technologically. As easily as we installed drain tiles, built steep-sided channels, raised levees, and constructed reservoirs, we can plug the drains, flatten the channels, cut the levees back, and resculpture our derelict agricultural and urban landscapes to resemble the old wetland topography.

Reengineering the site is just the first and the easiest step. Stimulating and nurturing the complex biological processes stunted and destroyed by the loss of natural wetlands will be far more difficult. Reestablishing wetland flora will require the elimination of tough and alien species that moved in to dominate the disturbed urban environment. These species must be contained, combatted, and managed. New species must be planted: seeds and plants brought in, tended, and nurtured until they are strong enough to stand on their own. Breeding grounds must be created for those rare species of fish that today we cherish but rarely see.

We must construct, through the selective placement of plants and materials, the appropriate niches that nature took centuries to create. Water levels must be adjusted to meet the living and breeding requirements of the wildlife we introduce. The unique relationship between land, plants, and water must be reestablished in order to purge the stream of pollutants. Virtually every aspect of the wetland environment will need monitoring, control, and experimentation. Some native species will not survive; others will thrive; some will quickly succeed on their own; others will need long-term protective management. The use of fire, an important element of the prairie ecology, is an example of the kind of management that will be required.

I am not describing a hypothetical scenario. I am describing what we are doing at the Des Plaines River Wetlands Demonstration Project, thirty-five miles north of Chicago in Wadsworth, Illinois. We are restoring, in this fashion, 450 acres of former wetland and the attendant three miles of river channel. The site, not surprisingly, was once a wetland. It was drained in the mid-1800s and used for farming until the early 1970s. Prior to the mid-1970s, gravel was excavated from the site, leaving three large pits scarring the landscape. The river, draining over two hundred square miles of agricultural and urban watershed, enters and leaves the site contaminated by soil particles, nutrients, and, often, luxuriant growths of algae (Hey and Philippi 1985). The project confronts all the problems noted above.

Since September 1989, six of the eight planned wetland areas have been reconstructed and put in operation. Already we are seeing results. The diversity of prairie and wetland plants and wildlife that once predominated is now returning. An ornithologist observed in 1990 a nearly 4,000-percent increase in the number of waterfowl using the site since 1985 (Hickman and Mosca 1991). This is very encouraging, considering that during the same five-year period the U.S. Fish and Wildlife Service reported a decrease nationally in the populations of most waterfowl. Several bird species that are endangered in Illinois also have been observed at the site, including the great egret, pied-billed grebe, and yellow-headed blackbird.

At the same time, the wetland areas that the river water passes through have been able to purge the stream of more than 80 percent of unwanted contaminates, while attenuating the movement of floodwaters downstream. Even a naked eye can easily observe the difference between the muddy, murky waters entering the wetlands and the clear water that returns to the river after traveling through the wetland. This is only the beginning of our research demonstrating how wetlands—with optimal design parameters defined by the research at this site—can be constructed as an alternative to existing water-treatment programs.

The project is intended not only to show that restoration is possible, but to teach us the best ways to re-create wetland environments. Its very large research component is intended to answer many of the questions that are being asked by re-

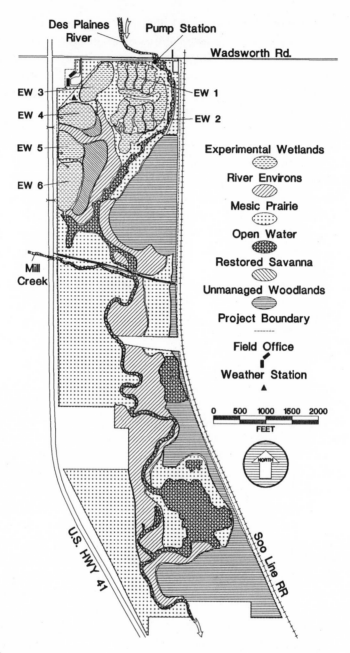

Figure 1   Des Plaines River Wetlands Demonstration Project, Wadsworth, Illinois

searchers and policy-makers alike: Can water quality be restored by filtering a stream through a single small wetland? What is the appropriate scale? To what extent can flood stages be reduced to prevent downstream damage? What is the most efficient location for wetland, flood storage? How rich and stable will created wildlife habitats be? Will the sites attract migratory waterfowl? How much will rehabilitation cost? In our living laboratory, in which teams of engineers and scientists from a number of different institutions work, we hope to develop the blueprint for enriching our urban landscapes and managing our urban waterways on a scale that could permanently enhance our urban environment.

At a symposium of the British Ecological Society, A. R. Clappen (1965) said:

> [W]e may frequently be faced with the need to refashion the lost ecosystems in as much detail as possible, so as to conserve elements of their natural flora and fauna or to extract from the ecological information and the scope for ecological information and the scope for ecological experimentation that their loss threatens to deny us. If this is our aim we have two main courses open to us: to allow a natural system to reconstitute itself through the stages of its primary and secondary succession—and successional time may be very long indeed; or to draw upon our ecological knowledge and practical skill in order to reduce the waiting period to a minimum.

We agree with Clappen that nature, left to its own devices, would have a very slow and difficult time reconstituting the systems that once existed. But if through our practical engineering skills we restore the physical conditions, and through our scientific knowledge we provide a boost to nature's handiwork, we feel confident that those systems can be restored. I speak not only for myself but for the team of botanists, microbiologists, mammalogists, fishery scientists, geologists, pedologists, historians, and hydrologists who are working together on the project when I say that it is a great adventure.

References

Bell, H. E. III. 1981. *Illinois wetlands: Their value and management.* Doc. no. 81-33. Illinois Institute of Natural Resources.

Changnon, S. A., Jr. 1985. Research agenda for floods to solve policy failure. *Journal of Water Resources Planning and Management* 3(1). American Society of Civil Engineers.

Clappen, A. R. 1965. Symposium address. In *Ecology and the industrial society.* New York: John Wiley.

Dahl, T. E. 1990. *Wetland losses in the United States: 1780s to 1980s.* Washington: U.S. Department of the Interior, Fish and Wildlife Service.

Gale, E. O. 1902. *Reminiscence of early Chicago and vicinity.* Chicago: Fleming H. Revell.

Goldstein, J. H. 1971. *Competition for wetlands in the Midwest: An economic analysis.* Resources for the Future, Inc. Baltimore: Johns Hopkins Press.

Hey, D. L., J. M. Stockdale, D. Dropp, and G. Wilhelm. 1982. *Creation of wetland habitats in Northeastern Illinois.* Doc. 0.82-09. Springfield, Illinois Department of Energy and Natural Resources.

Hey, D. L., and N. S. Philippi, eds. 1985. *Baseline survey of the Des Plaines River Wetlands Demonstration Project.* Volume 2. Chicago: Wetlands Research, Inc.

Hickman, S. C., and V. J. Mosca. 1991. *Improving habitat quality for migratory waterfowl and nesting birds: Assessing the effectiveness of the Des Plaines River Wetlands Demonstration Project.* Chicago: Wetlands Research, Inc.

Horton, A. H. 1914. *Water resources of Illinois.* Springfield: State of Illinois River and Lakes Commission.

Illinois Environmental Protection Agency. 1984. *Progress.* 11(1) Springfield.

National Research Council (U.S.) Committee on Restoration of Aquatic Ecosystems—Science, Technology, and Public Policy. 1992. *Restoration of Aquatic Ecosystems.* Washington: National Academy Press.

U.S. Environmental Protection Agency. 1984. *The cost of clean air and water.* Report to Congress. Washington: U.S. Environmental Protection Agency, Office of Policy Analysis.

Wooten, H. H., and L. A. Jones. 1955. The history of our drainage enterprises. In *The Yearbook of Agriculture 1955.* Washington: U.S. Department of Agriculture.

Lake Tahoe: A Microcosm for the Study of the Impact of Urbanization on Fragile Ecosystems

Charles R. Goldman

Introduction

The pioneering Illinois biologist Stephen A. Forbes, with truly remarkable insight, used the word *microcosm* to describe the dynamics of the lake ecosystem to the Peoria, Illinois, Historical Society. His classic 1887 paper based on this address, "The Lake as a Microcosm," is to this day required reading in many ecology and limnology courses throughout the world (Forbes 1887). At this early date Forbes already had a profound understanding of the complex interactions of the biota of lakes. What Forbes did not appear to appreciate in his eloquent zeal to describe the interdependencies of the plants and animals within the lake ecosystem was the total dependence of this system on the surrounding air and watersheds. Pollution was a less dominant factor than it is today, and lakes were certainly closer to being isolated microcosms than they are now or are likely ever to be in the future. To sustain the environmental quality of lakes and the many cities that are now growing on their margins will require an entirely new approach to both the planning and the environmental management of lake and city alike.

In the relatively small (eight hundred square kilometers) Lake Tahoe basin, pollution, in the sense of increased nutrient loading of the lake, has resulted largely from land disturbance and atmospheric deposition on the lake and watershed as snow, rain, and dust. Tahoe is the tenth-deepest lake in the world, with a surface area of five hundred square kilometers. Its large volume would be sufficient to cover the entire state of Texas to a depth of about eight inches! The urbanization of the Tahoe basin has resulted in a complex array of environmental, social, and political problems. Some of the conflicts have already been resolved, while others will doubtless continue into the next century. Reviewing the various problems while maintaining a focus on the lake itself may provide useful guidance in other similar situations, both foreign and domestic, where planners, in the face of increasing population pressures, hope to maintain environmental quality during the inevitable urbanization of areas adjacent to sensitive freshwater or marine ecosystems.

Lake Tahoe was first seen by the invaders of the western United States in 1844, when General John Fremont followed the Truckee River to its source and viewed this remarkably beautiful expanse of cobalt blue waters. Located in a large graben fault basin at an elevation of just under two thousand meters at the crest of the Sierra Nevada between California and Nevada, the lake was appreciated only by Paiute Indians in previous centuries. Its tranquillity, however, was to be short-lived. By 1870, the mines of the Comstock lode in Nevada required timber, and the relatively lush, virgin coniferous forest of the basin provided an easy source of logs. These could be floated in the lake to the outlet and then rafted down the Truckee River to Reno for wagon transport to the Virginia City mines. Still, the primitive mining practices of the day did not appear to cause the level of destruction that modern clear-cutting does, as rapid regrowth of brush and trees repaired most of the damage done. Lakes are truly reservoirs of history, in the sense that they accumulate, first in their waters and then in their sediments, a nearly indelible record of whatever has transpired on their watersheds (Goldman 1985). By examining the sediment record of Lake Tahoe, it has been possible to show conclusively that the post–World War II period created much more erosion and nutrient loading of the system than occurred during the Comstock mining era.

By the turn of the century, summer vacationing had begun on a limited scale and a few hardy individuals were overwintering in the basin. First stage coaches and then the steamer *Tahoe* provided transport within the basin, as the more affluent San Francisco population began to stake out lakeshore property at Tahoe for summer vacations. A small-gauge railroad came into general use, and the abundant cutthroat trout population provided the basis for a few years of commercial fishing. Still, use of the basin was largely limited to summer visitors until World War II ended and Tahoe became the site of new casino construction at the Nevada state line at both the north and south ends of the lake. Development was now underway in earnest, and concern for Tahoe's future stirred in the hearts and minds of residents and visitors alike. The remarkable clarity of Tahoe's waters was legendary ever since Mark Twain had described the crystal clear waters in his book *Roughing It*. Maintaining this unique water quality became the central focus of political action and controversy that has continued to this day.

The Struggle for Regulation

By the late 1960s there was ample evidence of the rising population in the basin, and with it came the inescapable realization that, somehow, development in the basin had to be brought under control. The author, along with officials of the League to Save Lake Tahoe, made visits to the governors of both states, Laxalt of Nevada and Reagan of California. Using aerial photographs of sediment plumes entering the lake and recounting the sad fate of other lakes in the United States and abroad, we were able to help convince both governors that a bistate agency was required to assure more orderly, controlled development in the Tahoe basin. In 1965 a joint study committee of the California and Nevada legislatures examined the feasibility of establishing a regional planning agency. After several years of negotiations, the Tahoe Regional Planning Agency (TRPA) was created in 1970, with the charge of regulating all further development in the basin.

In the beginning the agency had four members appointed by the states and six from the five counties that surround the basin. In 1971 Robert Bailey, in a joint report of the TRPA and the U.S. Forest Service included in the *Geology and Geomorphology of the Lake Tahoe Region: A Guide for Planning*, developed a land capability system that classified land into three major categories of high-, moderate-, and low-hazard lands (Bailey 1971). These could then be subdivided into fourteen reconnaissance-level geomorphic units from the most to the least capable of sustaining development. In 1982 the TRPA adopted a series of "threshold" values for various measures of water quality below which the lake should not be allowed to go. It soon became obvious that heavy local representation on the TRPA did not provide the intended protection of the resource since local political appointees found it difficult to vote against growth and commerce-promoting projects. For a short time, California even appointed its own regional planning agency, the CTRPA, to protect the California side of the lake. In 1980 the bistate compact was revised, and out-of-basin representation was increased to eight members (Sabatier and Pelkey 1989). An important phase of the political-legal saga involved the League to Save Lake Tahoe and the office of the California state attorney general, who brought suit against the TRPA for not adequately protecting the resource. In view of this case and the continued degradation of the resource, U.S. District Court Judge Edward Garcia found it necessary in 1984 to impose a moratorium on development (*State of California v. TRPA et al.* 1985), which was upheld in 1985 on appeal. After months of meetings, a consensus on future development was achieved, and a workable compromise was hammered out in 1986.

The taxpayers of California showed their concern for the lake by passing an $85 million bond act to buy back the most sensitive land in the basin, thereby protecting it from future development, and to fund erosion control. Two years later, Nevada voters passed a similar bill to acquire land for the public on the Nevada side of the lake. Since 1985, the California Tahoe Conservancy has authorized over $77 million in site improvement and land acquisitions totaling about five thousand acres. Federal Burton-Santini funds, the Soil Conservation Service, and the California State Water Resources Control Board have invested almost as much over the years.

The latest challenge to the TRPA's authority was a developer's lawsuit against the

State of California and the TRPA (*Kelly v. State of California et al.* 1990). The TRPA was using an "individual parcel evaluation system" (IPES), which was largely based on the Bailey land classification system. The basic idea was to prevent development of the steepest, most erodible slopes. After extensive testimony in this 1990 court case, the attorney general was successful in defending both the IPES procedure and the TRPA's right to designate what land could not be developed. On appeal, the Nevada Supreme Court upheld the validity of TRPA's regional plan on 3 July 1993.

Doubtless the TRPA will continue to face various future legal challenges. The prolonged drought that by 1992 had lowered the lake level 2.5 feet below its natural rim is creating problems for Reno, Nevada, a major downstream water user. Further, boats are no longer able to leave or enter many of Tahoe's marinas. The problem of dredging marinas and the channels necessary to reach deep water is becoming an increasing source of contention between the permitting agencies and the commercial interests. Unfortunately, it is impossible to generalize on the environmental impact of dredging, except to say that nutrients are always released to the surrounding lake. However, since the nutrient content of the sediments varies greatly from site to site, as does the ability of the dredgers to contain disturbed sediment within marinas or with plastic curtains, permitting requires particularly careful attention. Research on the impact of dredging and methods of mitigation is currently underway by the Tahoe Research Group. It is unfortunate for current applicants that there have been violations of some of the strict requirements contained in the dredging permits previously issued.

Water Quality and Sewage Treatment

There was little objective evidence for water quality deterioration in the early 1960s, but the eutrophication of many eastern lakes and Lake Washington in Seattle was attracting national attention. The great controversy over phosphate detergent was raging, and scientist and layman alike recognized that water quality in Tahoe was probably endangered. Development in the basin had been essentially unbridled, with the casinos acting as a magnet for both gamblers and the large number of employees needed to maintain the twenty-four-hour gaming and hotel-motel industry. The flattest lands adjacent to the lake were the first to be built upon. The wetland areas in particular, which form both wildlife habitat and the ecotone buffer between land and water, were favored for housing and commercial development. The wetlands had acted as sponges for nutrients from the watershed, and the loss of these marshy areas was increasing the nutrient loading of the lake. In the spring, when snowmelt swelled the runoff from the increasingly disturbed watershed, giant sediment plumes were visible in the lake for weeks at a time. From

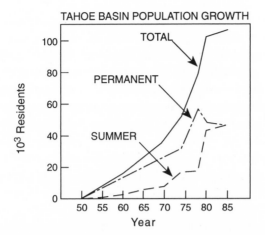

Figure 1 Human population growth in the Tahoe Basin from 1950 to 1985

an airplane, an increasing network of roads was visible extending to intensively developed areas well above the shoreline. Many of the roads were steep, and the bare roadsides were easily eroded.

By the early 1960s the small south Tahoe sewage-treatment plant was already being overloaded by the multiplying resident and visitor population (fig. 1), and overflow of effluent from the secondary treatment plant into the lake was occurring at more frequent intervals. In recognition that sewage was a major unsolved problem in the basin, with septic tanks and leach fields occupying an increasing portion of the flatter land, a short-lived but effective organization, the Lake Tahoe Area Council (LTAC), was formed in 1958 and succeeded in gaining a Fleishmann Foundation grant to study the various options for sewage disposal.

A San Francisco area consulting company, Engineering Science, was selected by the LTAC to undertake a study of alternative sewage-disposal options in the early 1960s. The author, then a young assistant professor at the University of California at Davis, was already studying Lake Tahoe as part of a study of northern California lakes funded by the National Science Foundation and now in its thirty-third year. I was asked by the LTAC Advisory Board to participate in the study as the consulting limnologist. Since most of the basin was still unsewered, the south Tahoe municipal sewage treatment plant was enlarged and extensively modified for tertiary treatment in an attempt to meet the increasing demands of new housing and casino construction. Tertiary treatment takes primary (settling out of solids) and secondary (bacterial digestion of organic material) treatment to a higher level by chemically precipitating out phosphorus and metal contaminants. Residual organics are then stripped out by activated charcoal, producing near drinking water quality effluent. It was even suggested that the effluent from this advanced wastewater-treatment plant was of sufficiently high quality that it could be released into Lake Tahoe without contributing to the eutrophication of the system. I strongly disagreed. Although the tertiary treatment system was excellent for phosphorus and heavy metal removal, the plant was not effective at nitrogen removal. In fact, when added to cultures of phytoplankton contained in the lake water, the discharge from the plant was extremely stimulating to algal growth. On this basis and with the support of a distinguished Swiss engineer, Karl Wuhrmann, the author was able to convince the LTAC consulting board to make the decision to export the treated sewage from the south end of the lake. Indian Creek Reservoir, which received the tertiary-treated effluent, had a number of fish kills resulting from the very high ammonia levels in the effluent. The ammonia formed highly toxic ammonium hydroxide during periods when extensive algal blooms developed in the reservoir. Because of the high cost of tertiary treatment and the necessity for export, the plant eventually downgraded to secondary treatment. A collection system and export line for untreated sewage was then constructed at the north shore to carry sewage from the basin to a new treatment plant in Martis Valley. The expanded capacity for sewage export from the basin stopped one source of pollution, but certainly made it possible to accelerate casino and motel expansion as well as further housing development. With the sewage problem under control, attention could be shifted to the many nonpoint sources of pollution to the lake.

Air Quality

Despite the fact that Tahoe is an ideal site for atmospheric inversions, air quality in the Tahoe basin was for many years considered a matter of little or no concern. In the early 1970s the author reported to a meeting of local science teachers that air pollution at a busy crossroads near the casinos was not greatly different from that being measured in metropolitan areas of California. The comparison brought on a furious and vindictive outcry in the local Tahoe press against the report of this observation which, incidentally, had been taken directly from a California Air Resources Board publication. It was rather like the fable of the emperor's new clothes. Everyone could see the smog developing, but few seemed willing to acknowledge that this near-pristine high-altitude lake basin could possibly have declining air quality.

Since air pollution could potentially limit growth in the basin, the early political response was to suggest that it was largely derived from outside the basin. A physicist at the University of California, Thomas Cahill, who specializes in identifying the sources of airborne pollutants, found that most of the pollutants, with the exception of sulfur, were largely generated within the basin. Once the "rose-colored glasses" were removed and the population began to acknowledge the existence of smog and the generally declining air quality in the basin, more attention was directed toward the study of this problem and various proposals have been made to reduce it.

It is important to recognize at this point that maintaining the casinos and the vacation-oriented commerce in the basin requires a very large number of in-basin vehicle-miles per year. Substantial fleets of diesel-powered supply trucks and tour buses from California cities shuttle loads of goods and gamblers twenty-four hours a day into and out of the basin. During winter months the buses often keep their engines running between loads rather than face the difficult cold starts. Further, the engines of out-of-basin vehicles are tuned for near-sea-level barometric conditions and thus produce extra pollutants in the thinner air at the higher elevation of the Tahoe basin.

The problem of air quality in the basin is intensified by the use of inefficient, highly polluting woodburning fireplaces, together with the strong atmospheric inversions already mentioned. Woodburning stoves and enclosed fireplaces, which burn more efficiently, are in greater usage now and are required in new construction. Eventually, even these may be restricted, to reduce emissions. Air traffic by commercial jet aircraft has increased markedly in recent years with the expansion of the South Tahoe airport. In addition to the escalating visitor-miles, there is also heavy automobile traffic in the basin, necessitated by a lack of home mail delivery and the large number of commuting employees associated with the gaming and recreation industries. Light rail has been recommended to reduce in-basin air pollution, but as yet, limited bus service is the only mass transport available.

What was particularly important from the standpoint of the management of air quality at the lake was the deployment of improved air-sampling equipment. An inexpensive portable air-sampling system was developed for use in the basin by Dr. Cahill. Another important innovation was the placement of monitoring equipment on spar buoys on the lake. This demonstrated conclusively the significance of atmospheric nutrient loading of the lake, which was highest near the population centers at the north and south ends of the lake and decreased toward the center of the lake. The importance of forest fires, even at great distances from the lake, in fertilizing the surface waters has also been demonstrated (Goldman et al. 1990).

Vegetation, Erosion, and Stream Transport

In considering the lake and its surrounding watershed as a single unit or "landscape," as many ecologists would now prefer to view it, the terrestrial vegetation is of particular concern. Trees not only serve the global ecosphere by providing oxygen and removing carbon dioxide, but they also help anchor the soil on the steep slopes and reduce the erosive impact of rain striking the ground. Further, and very importantly from the standpoint of Tahoe, they recycle the nutrients released by the soil microorganisms as well as nutrients contained in the rain and snow. One additional benefit of the extensive and largely second-growth coniferous forests of the basin is the filtration of the air mass itself. Atmospheric dust and airborne pollutants adhere to the pine and fir needles, or are sufficiently slowed in their transport that they fall out on the forest floor rather than on the lake surface. As noted earlier, wetlands near the mouths of the many streams serve the same general purpose of keeping or at least slowing the flow of nutrients from the land to the lake. All of these factors combined do much to retain and recycle nutrients, thus preventing them from contributing to the already increasing eutrophication, or fertility, of Lake Tahoe.

Proof of the importance of the terrestrial environment in buffering the lake's nutrient regime is provided by the stream surveys of the Lake Tahoe Interagency Monitoring Program. Six to ten streams in the basin have been regularly monitored for their nutrient transport to the lake. These streams provide about 50 percent of the total inflow to the lake and include the major tributary, the Upper Truckee River, which provides about 40 percent of the total stream inflow. The streams, unless they include fertilized lawns or highly disturbed areas in their watersheds, typically contain less nitrogen per unit volume than rainwater.

The extensive road network that has accompanied the general development of the watershed has been a major source of eroded sediments and associated nutrients. In particular, the phosphorus content of runoff is closely correlated with the sediment load. This has become particularly important as the nitrogen content of the lake has gradually built up over the last twenty years and the algae have grown increasingly sensitive to phosphorus inputs. The California Tahoe Conservancy, which manages a number of erosion control projects, has concentrated on reducing the erosive energy of roadside drains as well as stabilizing road cuts with rock work.

The ski slope at Heavenly Valley on the south shore of the lake bears stark testimony to the difficulty of reestablishing vegetation once it has been cleared. For the last thirty years, the barren, scarred slopes have remained unhealed despite more or less continuous attempts in recent years to vegetate them. Various kinds of native ground cover have been planted over the years, but it is extremely hard to reestablish ground cover on steep slopes at this elevation in a region where the winters are long and the summers very dry. Fortunately, progress is being made to improve the technology of establishing subalpine plants, and perhaps one day the ski run will no longer be so visible and the sediment loads its slope delivers to the lake will be reduced. Other ski areas in the basin have similar problems with soil erosion and require attention.

Some streams in the basin have suffered directly from development. The Third Creek channel in the Incline Village area of the north shore was moved during the development of a golf course, resulting in a large transport of sediment to Crystal Bay. Another serious and long-lasting problem developed when gravel mining was allowed in the Blackwood Creek channel. The resultant increase in stream velocity, which is typically dissipated in undisturbed stream channels, led to severe channel erosion and nitrogen yields from two to nine times that of the six other streams monitored at the time. The sediment discharge was so great that the shoreline filled in, leaving boat moorings and piers high and dry. Since it was a federal agency that allowed the gravel mining operation in the first place, it seems fitting that the federal government is now investing heavily in the rehabilitation of Black-wood Creek.

Lake Response to Nutrient Loading

The response of the lake to development in the basin as a whole became gradually evident from the data collected since 1959 (Goldman and Carter 1965). The primary productivity, or growth rate of the lake's natural algal population, was increasing at about 5 percent per year (Goldman 1981). Figure 2 illustrates the steady increase in productivity over the last three decades. There is a good deal of interannual variability. A senior systems analyst for the Tahoe Research Group, Evelyne de Amezaga, was the first to correlate the percentage change in productivity from the previous year with the nitrate discharge of the Tahoe tributaries. This important discovery clearly tied the nutrient input from the watershed to lake productivity. We were subsequently able to improve our understanding of this relationship by recognizing that these nutrients were accumulating in the deep waters of Tahoe, and that deep mixing of the lake when late-winter storms occurred during years of heavy runoff was largely responsible for the percentage change (Goldman and Jassby 1990).

The importance of having long-term data sets can scarcely be overemphasized. There is so much interannual variability in environmental data of almost any kind due to changing weather conditions that, in the case of Tahoe, five years of data were necessary to establish a significant change in the productivity of the lake, and six years to establish a significant loss of transparency (fig. 3). Figure 3 illustrates the loss in transparency of almost half a meter per year. If, however, only the drought years of the mid-1970s had been considered, one would have concluded that the lake was actually improving (fig. 4).

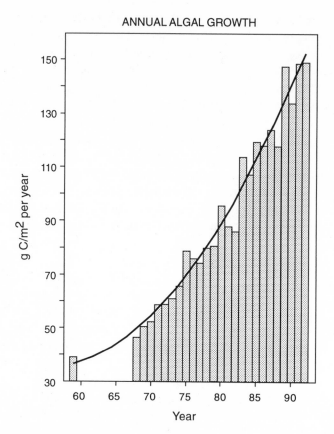

Figure 2   Annual integrated primary productivity as determined at the U.C. Davis Index Station, Lake Tahoe, using the ^{14}C method; 1968–1992. The 1959 value was determined at a comparable station near the east shore.

Lessons from the Tahoe Basin for Sustaining Environmental Quality

Despite over thirty years of development and the environmental conflict that accompanied it, Lake Tahoe is still in remarkably good condition. It owes this largely to its enormous volume, its relatively small and infertile Sierra Nevada watershed, and conservation efforts that were established early enough to make a difference. Although productivity has increased at over 5 percent per year, the clarity of the water still maintains Tahoe's position among the clearest large lakes in the world. In contrast to Lake Washington in the Seattle area, which experienced enormous eutrophication from the discharge of numerous sewage treatment plants and then showed rapid recovery thanks to a short retention time and the diversion of sewage to Puget Sound (Edmondson 1991), the sewage input to Tahoe was cut off early in the development stage, and shortly thereafter solid wastes were no longer buried within the drainage basin but instead exported to landfill sites outside of the Tahoe basin. Had this not been accomplished, Tahoe would have rapidly moved to a more productive, eutrophic state and lost forever its cobalt blue character.

Of great importance in the environmental saga of Lake Tahoe was the early acquisition of limnological research data that served to monitor conditions in the lake years in advance of the establishment of the interagency monitoring program. The high sensitivity of the carbon-14 method in measuring actual growth rates of algae in this ultraoligotrophic (extremely low in productivity) system made it possible to convince regulatory agencies and the general public that sewage diversion in itself was not sufficient to maintain the desired level of water quality in the system. It does not appear to have been an accidental correlation that the rising human population and the increasing algal growth rate mirrored each other so closely. Further, the research program enabled us to follow the changing biological dynamics of the lake, which included the introduction of the exotic opossum shrimp, *Mysis relicta*. Although added as fish food, the initial impact was to destroy a major food source for kokanee salmon, one of the important sports fish in the lake. Last, but certainly most critical from the standpoint of the ecology of the system, was recognition that the air, land, and water were inseparably linked, and

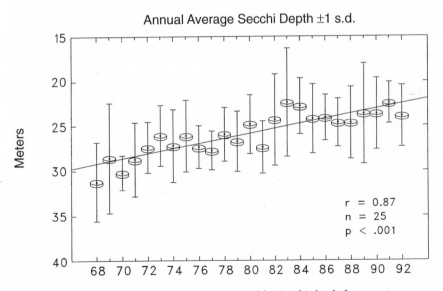

Figure 3 Transparency of Lake Tahoe as measured by Secchi depth from 1968 to 1992. Each point represents the average of about 35 measurements taken throughout the year.

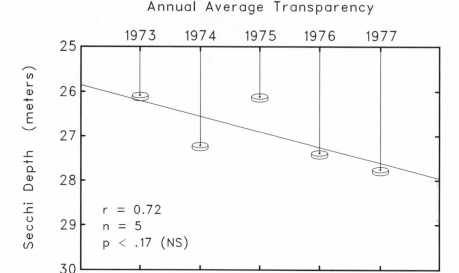

Figure 4   Temporary, drought-induced improvement in transparency at Lake Tahoe during the period 1973 through 1977

that deterioration of the landscape and airshed contributed significantly to the decline in water quality.

The political side of the Lake Tahoe story shows the importance of establishing an agency or agencies with real enforcement capability. For a number of years in the 1960s, pollution control in the Tahoe basin was essentially unenforced. Even the initial establishment of the bistate agency (TRPA) was insufficient, since local domination of the agency greatly reduced its effectiveness. However, increasing the out-of-basin membership on the TRPA board gave it both the will and the necessary strength to develop basinwide environmental controls. Finally, there was the existence of an activist organization that had the membership strength, the enthusiasm, and the political muscle to influence the course of government controls in the basin over a period of nearly three decades. New controversies continue to arise, and the ongoing education of the public on these issues is a principal hope for maintaining environmental quality in the Tahoe basin and in similar lake-city situations throughout the world.

References

Bailey, R. G. 1971. Geomorphic analysis of the Lake Tahoe region. In *Geology and morphology of the Lake Tahoe region: A guide for planning.* Prepared for the Tahoe Regional Planning Agency and Forest Service. Zephyr Cove, NV: U.S. Dept. of Agriculture.

Edmondson, W. T. 1991. *The Uses of Ecology.* Seattle: University of Washington Press.

Forbes, S. A. 1887. The lake as a microcosm. *Illinois Natural History Survey Bulletin* 15:537–50.

Goldman, C. R. 1981. Lake Tahoe: Two decades of change in a nitrogen deficient oligotrophic lake. Plenary lecture. *Verhandlungen, Internationale Vereinigung für Theoretische und Angewandte Limnologie.* 21:45–70.

———. 1985. Lake Tahoe: A microcosm for the study of change. Symposium proceedings: 125 Years of biological research, 1858–1983. *Illinois Natural History Survey Bulletin* 33:247–60.

Goldman, C. R., and R. C. Carter. 1965. An investigation by rapid Carbon-14 bioassay of factors affecting the cultural eutrophication of Lake Tahoe, California-Nevada. *Journal of Water Pollution Control Federation* 37:1044–1059.

Goldman, C. R., and A. D. Jassby. 1990. Spring mixing depth as a determinant of annual primary production in lakes, p. 125–132. In M. M. Tilzer and C. Serruya (eds.), *Large Lakes: Ecological Structure and Function,* NY: Springer-Verlag.

Goldman, C. R., A. D. Jassby, and E. de Amezaga. 1990. Forest fires, atmospheric deposition and primary productivity at Lake Tahoe, California-Nevada. *Verhandlungen, Internationale Vereinigung für Theoretische und Angewandte Limnologie* 24:499–503.

Kelly v. State of California et al. 1990. Ninth Judicial District, State of Nevada. Case no. 18325.

Sabatier, P., and N. Pelkey. 1989. *Land development at Lake Tahoe, 1960–1984.* Report to the Center for Environmental and Urban Problems, Florida Atlantic University. Fort Lauderdale.

State of California (People of) x ex rel. John K. Van de Kamp v. TRPA *et al.* 1985. (Ninth Circuit, 1985) 766F.2d 1308; 766F.2d 1310; 766F.2d 1319.

Wetlands in the Urban Landscape of

the United States

James A. Schmid

Had wetlands not been drained, filled, and developed, many American cities would look quite different today. Pierre Charles L'Enfant's plan replaced a swamp with the young nation's new capital city, Washington, D.C. Likewise, sections of Boston, Philadelphia, New Orleans, San Francisco, Chicago, Seattle, Dallas, Miami, Juneau, and New York, together with their suburbs, now display dry land where marshes and muddy woods once thrived. Urban progress long meant leveling hills and filling the wetlands below to build houses, highways, and airports, or to eliminate a source of mosquitoes, which were considered a public nuisance and health hazard.

To preserve wetlands in the cityscape for the sake of the public interest was unheard of during three hundred years of European settlement in North America. But during the past twenty years wetlands have been recognized as socially valuable, even more valuable in some instances than the benefits of further urban development. With this recognition has come governmental regulation of alterations that would fill wetlands to make them more suitable for urban uses. Some applications to alter wetlands have been denied by regulatory agencies at the federal, state, and local levels; many approvals have entailed modifications of the applicant's original plans in order to reduce or eliminate wetland impacts. Some permits have required as conditions of approval that replacement wetlands be created to offset unavoidable losses.

This major change in public attitudes toward wetlands began after the extent of presettlement wetlands in the United States had been reduced by half through human activity. Efforts to protect wetlands are but one manifestation of late twentieth-century Americans' growing concern for environmental quality. Conservationists now widely hold wetland protection to be essential as the human population grows and its technical capabilities to transform the landscape continue to increase (Conservation Foundation 1988). Wetlands in their natural state today are acknowledged as offering multiple values to the public at large. The newly recognized importance of wetlands applies not only in urban areas but also in the nation's farmlands and forests, where the greatest expanses of wetlands are found.

The process of wetland regulation in the United States has evolved into a complex enterprise where authority is fragmented among multiple political jurisdictions and agencies. To date wetland regulation has been subject to technical uncertainties, a lack of clear and consistent regulatory policies, and minimal, erratic enforcement. Despite a host of laws and regulations, wetland losses continue across the nation. Controversy and change surround the rules, and the regulatory process currently varies from state to state and from municipality to municipality. The likelihood of gaining ultimate approval to destroy wetlands for development (with or without mitigation) often seems independent of the actual values and functions of the wetlands in question. At present the outcome of specific wetland protection cases frequently is determined more by politics than by site conditions or science.

As the twentieth century draws to a close, it seems probable that most new suburban and urban development will be deflected from at least the wettest of the remaining wetlands. If Americans demand from their political system a steady course in wetland protection, then corridors of wetlands and watercourses will be preserved—corridors more extensive than those that remain today in the urban landscapes that were developed during the eighteenth, nineteenth, and early twentieth centuries. Where wetland losses are unavoidable, the sponsors of construction projects may be routinely required to create new wetlands nearby to replace those lost. Land values in uplands near urban centers are relatively high, so it is unlikely that public efforts to create new wetlands will occur there; opportunities to restore degraded wetlands effectively, however, often can be found near cities and suburbs and may attract public funding.

My long-term interest in urban vegetation and the factors that shape it led me twenty years ago into environmental consulting. During the past two decades I have been much involved in the regulation of wetlands in Megalopolis and the Middle Atlantic states. My discussion summarizes current wetland issues in the urban environment, first touching on their values and losses, then sketching the regulatory framework, permit process, and landscape effects. The complexity of defining a wetland is noted, a subject of intense controversy at present but fundamental to public understanding as well as to attempts at regulation on all governmental levels. Finally, I mention several problems in the current regulatory process, concluding with observations on major political and technical challenges for wetland protection in metropolitan areas.

Wetland Values and Losses

Wetlands formerly were lands valued by their owners essentially at the cost of converting them into dry areas where buildings, homesites, or farms could be operated. Most wetlands still are valued by their owners primarily in terms of their use for economic purposes. In contrast, wetlands today are valued by agency regulators representing the public as biologically productive habitats where unmanaged waters and plant communities can produce a permanent stream of public benefits, particularly for inhabitants of urban areas. The resulting conflict of values drives controversy into the courts and political arena.

Waterways and their attendant wetlands are assigned high economic value when integrated visually and functionally into urban landscapes (Tourbier and Westmacott 1992). Wetlands offer open space for recreation and places where floodwaters can spread harmlessly for cost-effective, temporary storage during storms. They contribute to the physical and chemical purification of the waters that flow through them, benefiting recreational lakes, fisheries, and public water supplies. They provide temporary or permanent homes not only to the common wildlife otherwise absent from unhospitable urban settings, but also to many declining, threatened, or endangered species of both plants and animals. As wild spaces, wetlands also provide a variety of intrinsic and human values that are not readily quantifiable (Nash 1978). There is no present consensus as to how to evaluate wetlands; analysts disagree as to whether any correlation exists between functional values and the duration of site wetness. The current and potential values (both functional and economic) of a wetland ecosystem are not determined solely by its size, but also by the landscape context in which it exists.

Wetland benefits typically accrue not only to the owners of the wetlands but also, generally in greater proportion, to the public at large. Hence, unless tied to attractive waterbodies, these benefits usually are not reflected by the marketplace in the appraised or assessed value of the property that contains the wetlands. Indeed, wetland property values often are lower than those for adjacent uplands, reflecting the hazards that their wetness may present to would-be users prior to draining or filling, and the extra cost associated with their improvement.

The economic cost of filling wetlands for urban uses today usually is far outweighed by the enhanced value of the land in its filled condition. Therefore, landowners universally have a financial incentive to encroach upon wetlands for many purposes. Moreover, restrictions at any government level on the use of wetlands are viewed as infringements on the rights of owners to use their land as they see fit. Because nearly three-quarters of the wetlands that remain in the United States today are in private ownership, it is not surprising that wetland regulation is controversial. The federal government owns only 13 percent of existing wetlands; state governments, 11 percent; and municipalities, 2 percent (USFWS 1990). Public agencies, like private owners, often chafe at the burdens of the regulatory process when proposing development projects in wetlands.

According to rough estimates of the extent of wetland habitats, Americans have succeeded in destroying about 117 million acres (53 percent) of their wetlands in two hundred years in the forty-eight conterminous states (table 1). Hence wetlands as a percentage of these states' total land surface have dropped from 11 percent to 5 percent. The large extent of wetlands in Alaska raises the wetland cover remaining today in the United States as a whole to about 11.9 percent of the land surface.

The percentage losses have been greatest in California (91 percent) and in the farmlands of the midcontinent (Ohio, 90 percent; Iowa, 89 percent; Indiana and Missouri, 87 percent; Illinois, 85 percent), with ten states having lost more than 70 percent and twenty-two states more than half of the wetlands present during the 1780s. The actual wetland acreage losses were greatest in Florida, Texas, Louisiana, and Arkansas: in each the loss exceeds 7 million acres. In Alaska the percentage loss of wetlands to date has been so small that the Bush administration sought formally to exempt that state from its "no net loss" of wetlands policy.

Intensive urban uses nationwide historically accounted for about 22 percent of the saltwater wetland losses and 6 percent of freshwater wetland losses (OTA 1984). During the period 1954 to 1974, more than 9 million acres of wetland losses nationwide were attributed to urban development. Hence urban uses are a substantial cause of wetland losses, although they affect a far smaller wetland acreage than agricultural activities do (Frayer et al. 1983). The extent of remaining wetlands varies sharply among states (table 1). Within many states the extent of wetlands varies even more dramatically among counties and municipalities, not to mention individual landholdings. Likewise, the political interest in wetland protection varies markedly among states, municipalities, and property owners. A panel of the National Academy of Sciences recently recommended that restoration be undertaken to provide a net increase of ten million acres of wetlands across the nation during the next two decades as part of an integrated strategy to restore surface water quality, and it identified federal government programs that could be used for this purpose (USNRC 1992).

Given the fragmentary knowledge of wetland systems and the slow pace of education in communicating wetland values, it is not surprising that recent government policies regarding wetlands are anything but consistent. For example, almost all of the wetlands in thousands of acres of speculative building projects that reverted to the federal government as a result of the savings and loan association failures of the late 1980s received no protection, but hastily were returned to the private sector (Frederick 1991). Many of these properties were in urban or suburban areas. Local drainage districts in the Midwest still levy taxes on landowners to raise the funds to make certain that drained wetlands stay drained. More than a few municipalities order wetland vegetation removed from stormwater

Table 1   Approximate extent of remaining wetlands and wetland losses in the United States, 1780s to 1980s. States are listed in order of 1980s wetland acreage (data from Dahl 1990). The actual extent of regulated wetlands could differ from the reported values, depending on the definition of wetlands utilized.

State	1980s Wetland Acres (000s)	1980s Wetlands as % of All Land	1780s Wetlands as % of All Land	% Loss of Wetlands 1780s–1980	Wetland Acres Lost 1780s–1980s (000s)	Rank by Acres Lost
AK	170,000	45.3	45.3	0.1	200	45
FL	11,038	29.5	54.2	46	9,287	1
LA	8,784	28.3	52.1	46	7,411	3
MN	8,700	16.2	28.0	42	6,370	6
TX	7,612	4.4	9.4	52	8,388	2
NC	5,690	16.9	33.0	49	5,400	9
MI	5,583	15.0	30.1	50	5,617	8
WI	5,331	14.8	27.3	46	4,468	13
GA	5,298	14.1	18.2	23	1,545	20
ME	5,199	24.5	30.4	20	1,261	23
SC	4,659	23.4	32.3	27	1,755	19
MS	4,067	13.3	32.3	59	5,805	7
AL	3,784	11.5	22.9	50	3,784	15
AR	2,764	8.1	29.0	72	7,085	4
ND	2,490	5.5	10.9	49	2,438	17
NE	1,906	3.9	5.9	35	1,005	26
SD	1,780	3.6	5.5	35	955	28
OR	1,394	2.2	3.6	38	868	29
IL	1,255	3.5	22.8	85	6,957	5
WY	1,250	2.0	3.2	38	750	31
VA	1,075	4.1	7.1	42	774	30
NY	1,025	3.2	8.1	60	1,537	21
CO	1,000	1.5	3.0	50	1,000	27
OK	950	2.1	6.4	67	1,893	18
WA	938	2.1	3.1	31	412	36
NJ	916	18.3	29.9	39	584	33
MT	840	0.9	1.2	27	307	40
TN	789	2.9	7.2	59	1,150	25
IN	751	3.2	24.1	87	4,849	10
MO	643	1.4	10.9	87	4,201	14
AZ	600	0.8	1.3	36	331	38
MA	588	11.1	15.5	28	230	44
UT	558	1.0	1.5	30	244	42
PA	499	1.7	3.9	56	628	32
OH	483	1.8	19.0	90	4,517	12

Table 1 *Continued*

State	1980s Wet-land Acres (000s)	1980s Wet-lands as % of All Land	1780s Wet-lands as % of All Land	% Loss of Wetlands 1780s–1980	Wetland Acres Lost 1780s–1980s (000s)	Rank by Acres Lost
NM	482	0.6	0.9	33	238	43
CA	454	0.4	4.9	91	4,546	11
MD	440	6.5	24.4	73	1,210	24
KS	435	0.8	1.6	48	406	37
IA	422	1.2	11.1	89	3,578	16
ID	386	0.7	1.6	56	491	35
KY	300	1.2	6.1	81	1,266	22
NV	236	0.3	0.7	52	251	41
DE	223	16.9	36.4	54	257	39
VT	220	3.6	5.5	35	121	46
NH	200	3.4	3.7	9	20	49
CT	173	5.4	20.9	74	497	34
WV	102	0.7	0.9	24	32	48
RI	65	8.4	13.2	37	38	47
HI	52	1.3	1.4	12	7	50
48 Conter-minous States	104,374	5.0	11.0	53	116,756	
Total US	274,426	11.9	17.3	30	116,962	

detention basins. Hence there are still many gaps in the political framework for wetland protection.

The Framework of Wetland Regulation

Wetland regulation, to a much greater degree than the public acquisition of wetlands for preservation purposes, generates opposition because it infringes upon the presumed right of landowners to eliminate wetness from their property. Such a right, however, does not exist independent of the public's right to protection from environmental harm. The direct contribution of wetlands to environmental protection is most visible in urbanized areas, as notably documented, for flood damage reduction in the Charles River watershed, a tributary to Boston Harbor (Doyle 1986).

During the 1960s, the states began to enact laws regulating the alteration or destruction of wetlands, beginning with the Jones Act in Massachusetts (1963). By 1990, wetland development for urban uses was being regulated to some extent under state laws in thirty states, although only about fifteen presently regulate both inland and coastal wetlands. The first wetlands receiving protection were along tidal estuaries. In New Jersey, for example, the Wetlands Act of 1970 suddenly halted two decades of rapid filling of coastal marshes for residential and other purposes. Inland wetlands in most of New Jersey, the most intensively urbanized state, were not regulated at the state level until the Freshwater Wetlands Protection Act took effect in 1988. Those states where wetlands are most stringently protected have relatively small proportions of their land areas in wetlands, and their wetlands are at risk primarily from urban and suburban development, rather than from the expansion of farming activities. The interior western states and some states in the Midwest lack wetland protection programs at the state level; these nonregulating states encompass about one quarter of the remaining wetlands in the forty-eight conterminous states.

Americans long sought not to preserve but to eliminate wetlands through government action. During the mid-nineteenth century, the federal government gave more than sixty million acres of public wetlands to fifteen states for conversion to private farmland (Salvesen 1990). Floodplain wetlands were often deemed to be convenient and cheap sites for the disposal of material dredged from navigable waters by public or private entities. Until 1985, federal agricultural policies intentionally rewarded farmers for converting wetlands to crop production. The regulation of wetlands converted to farm use or from farm use to urban development remains the subject of acrimonious debate (Thomas 1990), but the withdrawal of federal financial incentives for conversion of wetlands into farming uses is promoting wetland preservation (Lant 1993).

The partial direct regulatory protection of wetlands at the federal level was enacted as Section 404 of the Federal Water Pollution Control Act Amendments of 1972. This statute established a permit program for the placement of fill in wetlands (and other waters) administered by the Army Corps of Engineers nationwide. Certain rule-making and oversight authority was assigned to the Environmental Protection Agency (USEPA). The role of federal agencies in the wetland permit program was revised in the 1977 amendments to the Clean Water Act and currently is again under review by Congress. Normal, ongoing farming and forestry were exempted from the Section 404 permit program, but the further expansion of farming into wetlands and the conversion of still wet but cropped wetlands to urban and suburban uses were not exempted.

The Corps of Engineers was slow to implement its regulatory authority during the 1970s but increased its efforts during the 1980s. Substantial discretionary authority resides in Corps district offices, and local interpretations at the Corps and other federal agencies have led to regional differences in administration of the Section 404 fill-permit program. In 1977, Congress authorized delegation of the Section 404 program for nontidal waters to the states. To date only Michigan has accepted delegation of the federal program. Several states have parallel permit programs that more or less overlap with the federal program in response to their individual legislative mandates. Most of the state wetland statutes are more protective of wetlands than the Clean Water Act. At present there is little interest among the states (except for New Jersey) in assuming federal wetland regulatory authority (Davis 1991).

Some municipalities have attempted wetland regulation, either on their own initiative pursuant to state enabling legislation or through delegation of state permitting authority. At present, however, wetland regulation is primarily the responsibility of the federal and state levels of government, and most agency technical expertise is found among the federal and state personnel rather than at the municipal level. Municipalities in some states can supplement other governmental protection of wetlands where the locally scarce wetlands are too small to warrant state and federal concern, and a few have done so (see, for example, the Pennsylvania guidance to municipalities in PADER 1990 and sample ordinances in Kusler 1983).

Wetlands—or more precisely, proposed fills or other drastic alterations to wetlands in order to accommodate new uses of the land—are regulated in order to protect the public against harm. Such regulation exposes public agencies to the risk of making permit denials that in effect deprive landowners of some or all new economic uses they might want to make of their property. In general the state and federal courts have upheld regulatory decisions protecting wetlands (Want 1989). At some point, however, regulation to prevent harm can become so onerous that courts may consider a specific property as effectively "taken" from its owner for public benefit. Takings claims were raised in more than half of four hundred wetlands cases reported over a thirty-year period, but in very few cases were the takings claims sustained by the courts (Kusler and Myers 1990). Most governments and conservation groups lack the funds to pay full (development) market value and seldom purchase wetlands outright from their owners to ensure the preservation of long-standing public benefits. Hence the threshold of regulation deemed by the courts to require compensation in accordance with constitutional guarantees is closely watched by developers and conservationists alike.

Such concerns have in some instances inhibited municipal ordinances protecting wetlands, even though municipalities are primarily responsible for land use regulation, and are closest to the actual wetland benefits and losses on the ground.

Few municipalities have on staff or choose to retain experts in wetland analysis. Increasingly, municipalities are asking that the wetland boundaries proposed by consultants on site plans submitted by landowners be confirmed by state or federal agencies, because the presence of wetlands subject to regulation by higher levels of government obviously poses formidable constraints on site planning for parcels targeted for subdivision. In this way municipalities can contribute substantially to the implementation of wetland laws during the course of reviewing urban and suburban development proposals, even in places where they are specifically prohibited from regulating activities in wetlands directly (as in New Jersey).

Only in a few states of the Northeast are wetland delineation and regulation chiefly municipal functions at present (Salvesen 1990). An increasing trend among municipalities is the imposition of review fees on development applications sufficient for the municipality to retain outside professionals to review the work of project sponsors. In this way even small municipalities can obtain wetland expertise and other professional assistance when reviewing proposed projects, without incurring costs to existing taxpayers or increasing their permanent staff.

The present regulatory system is not efficient in arriving at decisions. It has not stopped wetland losses, which still are believed to range into the hundreds of thousands of acres annually, but it has slowed those losses substantially in many parts of the nation, particularly in urban and suburban areas. It still provides many opportunities for interagency rivalry as well as for evasion of the responsibility for enforcement and for escaping the financial risk of permit denials judged to be takings. The regulatory process is continually evolving, and changes can be expected during the next decade.

Permits for Construction Activities in Wetlands

Except in those states that have published detailed, large-scale maps of regulated (usually tidal) wetlands, landowners are responsible for identifying the extent of wetlands on their properties and for avoiding such areas during new construction. That means establishing lines on the ground where wetlands give way to uplands, and then transferring those lines to property maps and engineering drawings.

Most boundary determinations, especially on large properties, are made by consultants working for the landowner at the time a property is being sold or a project is being planned. Agencies may confirm the proposed regulatory boundaries and may revise them on the basis of additional field analysis or reinterpretation of definitional guidance. It is essential to the orderly process of equitable land sales and development that reliable boundaries for regulated wetlands be established and then respected during project planning. Over the past decade wetland boundaries have fluctuated in response to regulatory definitions that have changed within the developmental time frame of construction projects, causing repeated redesign in some instances. The regulatory flux of the past twenty years has been far from optimal for either wetland protection or cost-effective land development.

Where it is not possible to avoid working in regulated wetlands, project sponsors are required to secure permit approval prior to undertaking regulated activities. Certain classes of minor activities in wetlands and waterways have been authorized by the Army Corps of Engineers through nationwide permits that, if approved by the state with respect to water quality protection, allow work without an individual application (some nationwide permits require agency notification before the work can proceed). Nationally or locally applicable conditions pertaining to the nationwide permit apply to the authorized activities. Corps permit regulations are revised every few years; current regulations are found at 33 *Code of Federal Regulations* 220–330.

Some states have statewide general permits that resemble the Corps of Engineers nationwide permits to a greater or lesser degree. They also may require notification of the state agency and confirmation of their applicability prior to use for specific construction activities. Some statewide permits require the permittee to compensate for wetlands filled pursuant to the permit. General permits are helpful to developers because they establish clear limits for what is allowed as a "minor" wetland encroachment to which they are entitled if they meet applicable conditions; for agencies such permits impose relatively little burden of staff review. A major concern of conservationists is that many "minor" wetland encroachments, if not mitigated, can add up to major, continuing wetland losses. Moreover, sites not inspected in the field may or may not in fact qualify for the requested permit, especially if the regulatory boundary has not been established accurately.

Wetland fill projects that do not fit within one or more general permits require an individual Corps of Engineers permit application triggering public notice and opportunity for review by various agencies. For activities deemed not to be water-dependent, the applicant must establish that no practicable alternative exists that would avoid or reduce the proposed fill in wetlands. The question of alternatives often generates protracted argument, and there are no clear standards by which to demonstrate that no practicable upland alternative site exists. (Unless the state approves the water quality certificate for the project, an individual Corps permit is not valid. Thus, even states without formal statutes for wetland protection can block federal fill permits if they choose to do so.) The Corps of Engineers performs a public interest review that considers many aspects and consequences of the proposed construction activity, including wetland protection. The review process for a federal-state, individual wetland permit generally is slow, and it imposes on applicants substantial front-end costs that are not necessarily in proportion to the wetland values at risk.

Approvals of those projects deemed to warrant the filling of wetlands now generally are being conditioned upon the attempted replacement of wetland values lost. Such conditions can be expensive for permittees to implement, especially in urban areas with high land values. Sometimes the replacement wetlands simply are not built, even though the wetlands are filled. Agency efforts to monitor wetland mitigation attempts long were minimal, but recent permits may include requirements that the applicant report annually on the success of mitigation efforts over a period of years. Performance bonds can be demanded by the Corps of Engineers to guarantee required mitigation of wetland losses, but seldom are.

Landscape Consequences

Wetlands and waterways occupy the topographically lowest sections of every urban landscape. They have been encroached upon to a greater or lesser extent in each American municipality to accommodate human activities.

Agencies increasingly demand extensive documentation of the need for a proposed wetland fill and the lack of practicable alternatives for shifting the proposed activity to an upland location. These requirements provide a strong incentive for project designers to avoid wetlands wherever possible, even in urban areas. If sustained over a long period of time, the avoidance of wetlands (combined with restrictions on stream enclosures that typically are based on flood protection and water quality statutes) will result in the widespread preservation of both isolated wetlands and wetland greenbelts along stream corridors in the urbanizing landscapes of the late twentieth and twenty-first centuries. To the extent that wetlands and other green spaces are preserved, both wildlife populations and human opportunities for passive recreation will be more diverse and abundant in future suburbs than in the dense, older, urban neighborhoods where today there is little unpaved land.

When unavoidable wetland encroachment is authorized, compensation for the wetlands lost typically is required. If sufficient new wetlands are created from dry land to offset wetlands lost, the future extent of wetlands would presumably remain stable or increase. The current federal standard generally is a 1:1 acreage replacement of wetlands lost. Some state regulations impose a 2:1 or greater ratio. The restoration or enhancement of existing wetlands sometimes is authorized in an effort to balance functional values even as the total wetland acreage is allowed to decrease. "No net loss" of wetlands is a much-publicized policy goal for the 1990s that can be achieved only through the technically successful creation or enhancement of new wetlands, given the persistent need for wetland alteration for many purposes.

Outside the regulatory process, wetland preservation and enhancement efforts can be anticipated chiefly along rivers and estuaries. In many metropolitan areas, the functions of public and private wetlands have been impaired by diking, ditching, or other historic alterations, and some such wetlands offer the opportunity for restoration to a full array of functions at relatively low cost. Whatever public funds become available for urban wetland improvement are likely to be spent to restore degraded wetlands along waterways. Innovative efforts to integrate waterways and wetlands into urban and suburban landscapes have been surveyed by Tourbier and Westmacott (1992).

Wetland Definitions

In order to regulate wetlands, legislators and administrative agencies first must define these distinctive, scarce, but widely distributed places where the land meets the water. Many scientists familiar with wetlands believe that the presence and limits of a wetland are scientific, technical matters driven by environmental facts, whereas the approval or disapproval of a construction permit is an administrative, policy-driven decision influenced but not dictated solely by wetland conditions.

Generally there is little difficulty in reaching consensus as to the wetland nature of lands that have water standing on them for long periods of time together with plants that exclusively inhabit soggy soils. Tidal marshes, for example, are relatively easily identified, inasmuch as their distinctive plant communities usually terminate where the land rises above the influence of regular tidal inundation. The first efforts to regulate wetlands focused on tidal wetlands, where definitions and jurisdictional boundaries raised little controversy. Inland, mucky bald-cypress swamps, peat bogs, cattail marshes, and red maple swamps with sedge tussocks and long-standing surface water seldom occasion disputes. Everyone recognizes them as wetlands, at least during wet seasons.

Difficulties arise today, however, when the boundaries of less obvious wetlands must be established in the field for purposes of regulation. Where state or local agencies have published maps of their jurisdictions, the limits of regulated wetlands are found where the maps say they are. But the detailed mapping of wetlands over wide areas is expensive, so determinations usually are made following site analysis on each specific parcel of land. Agencies generally seek to place the burden of the detailed site analysis on the landowner, who is the party most likely to benefit financially from changes in the status quo.

It is no small task to describe characteristics that regulated wetlands uniformly must display across so large and diverse a nation as the United States. Yet this step is fundamental if there is to be federal regulation. There is no societal consensus at present on just how extensive the lands regulated as wetlands should be. The federal definition of Section 404 wetlands used by the Corps of Engineers and USEPA for nearly fifteen years is this: "Those areas that are inundated or saturated by surface or groundwater at a frequency and duration sufficient to support, and that under normal circumstances do support, a prevalence of vegetation typically adapted for life in saturated soil conditions. Wetlands generally include swamps, marshes, bogs, and similar areas" (33 *Code of Federal Regulations* [CFR] 328.3; 40 CFR 230.3).

A major technical and political challenge is to apply this rather circular definition to the diverse array of wetlands across the United States. The federal agencies have focused on field indicators for three parameters, hydrology, vegetation, and soils, and have found ever-changing technical nuances leading into the frontiers of scientific knowledge. For regulatory purposes, each point on the ground must be recognizable as either wetland or not, whatever the combination of physical or biological features present.

In a wood or marsh where water can be observed above, at, or just below the ground surface for long periods of time, there is little disagreement that a regulated wetland exists. Many commonly acknowledged wetlands, however, typically are not inundated or saturated at all times. Moreover, the ecosystem of even a ponded wetland typically does not stop abruptly at the edge of the ponded water. A short-term field inspection during a dry season or rainless period may not document the presence of water on or near the surface of a wetland, however obvious that wetland may appear to the skilled analyst.

Careful attention to topography, slope, and soil conditions affecting the flow of water under typical regional circumstances allows inference of the extent of wetland hydrology in the field. Drift lines and watermarks can suggest the height reached by floodwaters, but useful, year-round field indicators of prolonged seasonal saturation or inundation are scarce. Hence field attention often focuses on soils and vegetation as persistent and recognizable clues to the presence of wetland hydrology. The precise length of time that water must be present in or above the surface soil to warrant regulation is a subject of active controversy at present; for this purpose vegetation and soils may or may not prove adequate indicators.

Plant communities are the most apparent wetland indicators. They are the most significant determinants of wetland functional values such as wildlife habitats and potential enhancement of water quality. Many plants also can be recorded on aerial photographs and observed readily throughout much of the year. Based on the analysis of aerial photographs, the National Wetland Inventory (NWI) maps sponsored by the U.S. Fish and Wildlife Service and prepared as overlays to U.S. Geological Survey topographic quadrangles provide a general guide to the location of prominent wetlands for much of the United States. The precision and cartographic accuracy of the NWI mapping varies regionally, and it seldom is adequate for regulatory purposes on specific tracts of land, especially in urbanized areas.

Many plants that virtually require saturated soil conditions to compete successfully in the wild ("obligate hydrophytes") do not require those saturated conditions throughout the year. Most can tolerate occasional periods of soil dryness. Robust stands of such plants typically suggest that the land is wet for prolonged periods and provide strong evidence for the minimum extent of any wetland boundary. Those regulatory definitions of wetlands based on plants alone generally focus on the species now characterized as obligate hydrophytes.

Many more kinds of plants, however, not only have roots capable of growing in the oxygen-poor conditions of saturated soils but also can compete successfully in well-drained uplands. Thus the mere presence of stands of these species ("facultative hydrophytes") may not be conclusive for wetland determination. Many of these broadly tolerant plants are among the most prominent components of the vegetation, particularly along the upslope margins of easily recognized wetlands. Hence considerable effort was devoted by federal agency personnel during the 1980s to categorizing the frequency of occurrence of hydrophytes in the presumed wetlands generally associated with these plants in each geographical region (Reed 1988). In New Jersey, for example, there are about three thousand species of higher plants that grow without cultivation. Of these 20 percent are obligate hydrophytes, 46 percent are facultative hydrophytes, and 34 percent are obligate upland plants considered to be generally intolerant of wetland conditions (Schmid 1990).

The NWI categories represent a consensus on a "best guess" concerning the proportion of individuals in a species that are growing under wetland and upland conditions in large geographic regions of the United States. Some state agencies, such as the Maryland Department of Natural Resources and the New Jersey Pinelands Commission, have adopted alternative indicator status designations within their jurisdictions, assigning a higher wetness ranking to many prominent species, but also reducing the ranking of other species from that established by NWI for the region as a whole.

The regional classification of a plant species, of course, is only a general guide for specific parcels of land. Any plant species documented locally as commonly growing in areas known to be ponded or saturated for more than 10 percent of the growing season is a hydrophyte (EL 1987), whatever its regional classification. Conversely, any individual plant growing in an upland is not a wetland plant, whatever its species classification as reported by NWI.

The species composition of plant communities intergrades along the local topographic moisture gradient from wet to dry. Thus a greater degree of reliability in regulatory boundaries often can be achieved if soil conditions are considered along with plant communities.

Soil morphology responds to the range of wetness conditions from ever-saturated or ponded, to less frequently saturated, to wet only during and immediately following precipitation events. The colors and textures of soil layers give considerable insight into the likely presence and duration of water in the plant root zone, reflecting the chemistry of oxygen, iron, and other elements in the soil solution under the localized moisture regime. The precise length of time that water must be present at various depths in the soil during the growing season to qualify as "hydric"—and thus as a defining characteristic for regulated wetlands—has varied over the past few years and currently is the subject of technical debate. Some very wet habitats with typical wet plants may never experience anaerobic conditions during the growing season because the soil water moves sufficiently to maintain free oxygen or the soil is too cold or too poor in nutrients to support bacteria. Also, where there is too little iron or manganese to provide tell-tale stains on the sand grains, the appearance of a soil may not reflect its protracted wetness.

Most soil morphologic features have the advantage of being available for examination all year long, and the National Technical Committee on Hydric Soils has formalized the technical definition for named soil series that can be categorized as hydric (NTCHS 1990). Unfortunately, the current NTCHS definition is not capable of field application to a specific pedon (soil sample), as opposed to a conceptually defined soil series for which long-term data are available. Any soil observed to be ponded or inundated for more than a week during the growing season is by current definition hydric, whatever its internal appearance. Significant soil conditions are not confined to those visible at the surface, but include features that must be sampled vertically at selected locations. Then conclusions derived from sample points must be extrapolated to the surrounding landscape.

Reasonably accurate, generalized information on the nature and distribution of soil types is available for much of the United States, at least in rural and outer fringe suburban areas. Urban areas generally have been little studied by agriculturally oriented soil scientists. The catchall "Urban Land" soil category presented in county surveys of metropolitan areas may provide no information on localized soil characteristics such as texture, color, and drainage. Undrained hydric soil conditions must actually be confirmed as present on the land, not merely shown on a (nonregulatory) county soil map, for a soil to be identified as hydric and thus potentially regulated by the Corps of Engineers as a wetland on a specific parcel. Many areas of hydric soil are encountered as "inclusions" in areas mapped as nonhydric series, and the converse also is true. (As much as 15 percent of the surface area of a mapped type can consist of inclusions without the need for a composite name to be assigned to the map unit.) County lists of nonhydric series likely to exhibit localized inclusions of hydric soils are available. Some nonfederal regulatory definitions of wetlands rely only on soil conditions; for example, Connecticut inland wetlands are defined by county soil survey maps (Aurelia 1988).

Where soil changes are abrupt, they may provide a defensible basis for drawing a boundary within a stand of vegetation that is potentially wetland but whose boundaries are otherwise uncertain. Like vegetation, however, soil conditions can be "borderline" or intergrade over considerable distances and thus also fail to define a clear boundary on specific parcels. Indeed, soil conditions are likely to be borderline exactly where vegetation conditions also are borderline, because both plants and soils respond to a seasonally fluctuating water table. Such problem areas typically display year-to-year variability in wetness as well.

Further adding to the difficulty of interpreting soils, some key characteristics defining hydric soil (e.g., gray colors, mottling, peat) can persist for many years after an area has been drained effectively while others, such as sulfidic odor or ferrous iron, may or may not persist throughout a single growing season. Although hydric soil characteristics alone may not demonstrate current wetness at a particular location, in combination with actively growing, hydrophytic plants, they offer strong evidence for wetland conditions.

Reliable, persistent field indicators of effective soil drainage, like those for current wetness, are scarce. Thus it is not always easy to document the current effectiveness of past drainage activities on lands where hydric soil colors and facultative hydrophytes are observed. There is considerable variation in the ultimate conclusion reached and in the level of proof required by regulators when establishing wetland boundaries in borderline cases.

Difficulties in defining individual wetlands usually occur where the soil, vegetation, and hydrologic features do not change abruptly or in close proximity to one another along a topographic boundary. Where a landscape grades imperceptibly from wetlands to uplands or there is a mosaic of small patches of wetlands and uplands, opportunities arise for substantial disparities in consultant and agency determinations of wetland boundaries.

Differences of a few feet in the placement of a wetland boundary have little significance when determining the extent of wetlands nationwide. A few feet in the siting of a wetland boundary, however, may exert major influence on what landowners are or are not allowed to do within a specific urban property. Permits to alter wetlands may or may not be issued. Upland buffer zones adjacent to the wetlands may be regulated by states or municipalities, further reducing "useful" land. The extent of regulated wetlands to be disturbed also generally determines the costs of compliance with permit conditions mandating the replacement of lost wetlands when their destruction is judged to be unavoidable. Thus there is considerable incentive for differences of opinion among participants to arise in specific cases involving all but the most obvious of boundary determinations for regulated wetlands.

Several federal agencies developed and refined the technical definitions of wetlands during the 1980s for use when carrying out their various responsibilities nationwide. The U.S. Fish and Wildlife Service (Department of the Interior) has the longest experience in identifying regional wetlands because of its statutory mandates to manage habitats for fisheries and waterfowl as well as other biota. The Soil Conservation Service (Department of Agriculture) was directed by the Food Security Act of 1985 to remove federal monetary incentives for farmers to convert wetlands to agricultural use, after many decades of primary effort directed to expediting drainage. The objective of the Corps of Engineers and USEPA is to standardize the methodology for determining the limits of wetlands regulated under the Section 404 permit program. The federal agency wetland manuals underwent several revisions and considerable field testing. Their efforts did not lead toward a convergence of definitions.

After interagency review, a federal manual for field determination of wetland boundaries was agreed upon jointly in 1989 by representatives of the Corps of Engineers, the USEPA, the Fish and Wildlife Service, and the Soil Conservation Service (FICWD 1989). Field indicators of vegetation, soil, and hydrologic conditions were to be recorded and combined in a systematic way to produce a wetland/nonwetland conclusion at pairs of boundary points, which then could be joined to form a regulatory line on individual parcels. A key provision of the 1989 manual was that wetlands could be regulated where both hydrophytic plants and (undrained) hydric soils were present, even in the absence of conclusive, direct observations of the presence of surface or subsurface water for prolonged periods. The joint federal manual was used primarily by the Corps of Engineers and the USEPA. Some states accepted the 1989 manual as the basis for determining wetland boundaries regulated under state law, and some states were authorized by Corps districts to establish joint federal and state regulatory boundaries on specific tracts of land. Uniformity in wetland definition has the obvious advantage of avoiding multiple regulatory boundaries on a single wetland tract.

In response to landowner complaints, the joint federal manual was withdrawn in 1991 from use by the Corps of Engineers through the 1992 Energy and Water Development Appropriations Act. Considerable federal agency effort was devoted to technical revisions of the 1989 manual. A substitute manual that eventually was proposed by the USEPA following modification by the Office of Management and Budget and by the Council on Competitiveness (56 Federal Register 157:40445–40480, 14 August 1991) drew intense criticism because of its lack of scientific basis, its difficulty of field application, and nonreproducible results (e.g., Bedford et al. 1992). The 1991 manual would have required specific field evidence of protracted wetness, a potentially major burden of data collection. The task of issuing a revised wetland manual was not completed during 1992 but left to the new administration.

Thus, at present the states of New Jersey and Pennsylvania rely on the 1989 joint federal manual for wetland determination under state law, but the five Corps of Engineers districts that also regulate wetlands in those states may use the 1987 Corps manual (EL 1987). Pennsylvania applicants currently are directed to provide double documentation where boundaries under the two manuals differ. Likewise, several New England states officially or informally use the 1989 manual. The resulting state-regulated wetlands may be larger than the current Corps-regulated wetlands on any given property, a situation confusing to everyone. Whether the revised federal manual that eventually emerges will, like the 1991 USEPA document, seek to intermix policy with science remains to be seen. It is to be hoped that a workable federal manual acceptable to the states will be adopted. The Corps of Engineers plans to implement nationwide a program for the training and certification of wetland delineators for work in individual districts beginning in 1994.

In the experience of this author, from the early 1970s until the mid-1980s, consensus on the extent of wetlands on specific project sites was swiftly reached with regulatory agency personnel following a close review of site conditions. During that period, the primary emphasis was on vegetation, with supplemental reliance on topography and soils. More recently, as use of the "three-parameter" federal manuals has become more widespread, technical agreement on the extent of wetlands has become less predictable and more subject to interoffice variation within and among agencies.

Several factors give rise to the growing disagreement over wetland boundaries. First, various agencies, and offices and individuals within those agencies, differ as to their intent to apply the vague federal wetland definition broadly or narrowly. One Corps staffer recently said his office was looking to throw out from jurisdiction all areas except those that could in no way ever be construed as uplands, while another field representative from a different Corps office said he was trying to protect every inch of hydric soil he could find.

Such differences in outlook are not confined to regulators. Competing consultants interpret wetland definitions at the margin on specific tracts of land. It is common practice for landowners to hire several consultants to perform independent wetland boundary analyses. After reviewing the several efforts to apply the same rules, the landowner then selects the most favorable results for submission to the regulatory agencies. In one instance in southeastern Pennsylvania, consultants working for the buyer and for the seller came to quite dissimilar conclusions regarding a seventy-five-acre property. The seller's consultant found about five acres of wetlands; the buyer's consultant, thirty-five acres. Both determinations were accompanied by extensive site documentation. The results of real estate appraisals based on the two wetland boundaries in turn diverged sharply. It is not possible that both consultants were correctly applying the same methodology, even if their point-specific data on site conditions were reported accurately.

The current rules are sufficiently ambiguous and the real world sufficiently complex that different conclusions drawn from the same data are possible, even when the investigators are examining the same evidence in the field at the same time using the same methodology. This, of course, was precisely the situation that the three-parameter federal methodology was intended to avoid.

Second, and probably more important, the technical refinement of relevant information necessary for wetland determination means that wetlands can be misdiagnosed by field personnel who lack adequate expertise to recognize hydrophytic plants and hydric soils according to the directives of the pertinent manual. When hundreds of plant species grow on a property of several acres, considerable effort by a competent botanist is necessary if they are to be cataloged accurately. Soil conditions must be examined at the right depth. Careful scrutiny also is essential to distinguish between gleyed horizons (probably hydric) and podzolic eluvial horizons (probably nonhydric) when undrained gray soils are examined.

Recent examples will illustrate the current situation. On a residential lot on a floodplain in southeastern Pennsylvania, a developer's engineer said there were no wetlands and provided a brief summary of site conditions (sixteen species of facultative hydrophytes, no hydric soils). No agency confirmation was sought. After the absence of wetlands was questioned during a municipal review of the project, the township retained a consultant to inspect the site. This second consultant agreed that there were no wetlands in the one-acre lot. The second consultant's very sparse data on site conditions, however, contradicted those of the first consultant by claiming two areas with hydric soil. The second consultant mentioned only "upland vegetation" (no species named). When a third consultant examined the site for neighboring landowners, it was found to include three species of obligate hydrophytes, eight species of wet facultative hydrophytes (eighty species of plants altogether), widespread hydric soil, and water standing ponded at the surface for many weeks during the spring in and around the site of the proposed house. The federal enforcement representative concurred that wetlands were present after site inspection and a review of the third consultant's thirty-page technical report. In another case in central New Jersey, the buyer's consultant found a nineteen-acre farmland site to be 63 percent wetlands (twelve acres); the seller's consultant, 16 percent wetlands (three acres). In New Jersey, unlike Pennsylvania, there is a formal state process (including full public notice and opportunity for public comment) for boundary verification, upon which a landowner can rely for at least five years. Cursory field inspection is no substitute for thorough investigation when identifying wetlands.

Problems in Implementing Wetland Protection

The current framework for protecting wetlands in the United States unquestionably is having an effect on the urban and suburban landscape. Duplicative regulatory effort at several governmental levels may not be an effective use of regulatory resources, and may drive up the cost of new construction without commensurate environmental benefits. Yet even overlapping regulatory authority may not protect all wetlands, as the ongoing annual losses nationwide of hundreds of thousands of acres demonstrate.

As noted previously, the rules for the determination of wetland boundaries have been in flux for more than a decade, giving rise to much of the uncertainty and acrimony between preservationists and development interests concerning wetland definitions. What might be expected by now to have become a routine technical procedure for boundary establishment is still a hotly debated political issue. A consensus has not been reached on how much land Americans want to label and protect as wetlands.

In many construction projects, wetland definitions change during the course of project planning. Project opponents often question wetland boundaries, because the complex methodologies for wetland determination may yield boundaries whose basis is not obvious on the landscape. Where wetland boundaries have not been confirmed formally, agency representatives can and do shut down construction activities that conscientious landowners thought were far outside regulated areas. Such regulatory intervention, however, is totally unpredictable with reference to any specific project, leading many to question the fairness of the wetland regulatory process and some to ignore it altogether with little fear of significant penalties.

Over the past fifteen years this author has noted many more instances where state and federal agencies declined to regulate wetland development than instances where regulation was extended into uplands. Nevertheless, small landowners may accept excessive regulation rather than risk the time and resources necessary to challenge an agency's boundary determination, even when it is technically weak.

Wealthy landowners often can defer wetland enforcement actions indefinitely through well-financed legal appeals and the assistance of agency inertia. Fines for wetland violations may be smaller than the cost to file a permit application.

Enforcement in many agencies has been assigned a lower priority than permit processing, with a resulting gap between what the regulations suggest and what actually is accomplished in wetland protection. Both federal and state agencies have considerable powers of civil and criminal enforcement against wetland violators. A few large fines and even jail sentences for wetland violations have been awarded by judges and juries. But not many. The absence or erratic enforcement of wetland protection requirements leads to frustration not only among preservationists but also among those members of the regulated community who go to great effort and expense to achieve compliance with permit requirements. Nonenforcement is serious because it allows a flow of economic rewards to those who damage wetlands outside the regulatory process while penalizing those who comply. It is not unusual for consultants or agency representatives making a boundary determination on one property to identify unauthorized fills on an adjacent property. Enforcement follow-up may or may not ensue.

In the experience of this author, the relevant technical issues generally receive at least cursory attention during the agency review process when a permit is recognized as necessary. Applicants often complain about the inordinate length of time required to secure even general permits for small fills. But technical mistakes sometimes are made.

In one application to fill along several hundred feet of the margin of a usually dry river channel in a Los Angeles suburb, the reviewing agencies long focused on nonissues—wetlands and an endangered fish that simply did not exist in the vicinity—without actually looking at the site. The fill was proposed near the midpoint of an urbanized, fifteen-mile long, dry stretch of river channel. No wetlands were present in the coarse alluvial stream bed deposits at the fill site. No fish of any kind could dwell here except for a few hours during rare, major flood events spaced many years apart. Numerous recent fills could be observed in the permanently flowing headwater section of the river, however. These obviously had not been regulated, and the endangered fish's habitat there had indeed been reduced. At the same time, miles downstream where the river again became permanent at the surface of the stream bed and the endangered fish was known to exist, active fill was spilling directly into wetlands and waters from the parking lot expansion at a large amusement complex. It also was not being regulated. Yet for years much agency concern ostensibly was expressed about the proposal to extend an existing embankment to meet an existing highway bridge abutment and fill a small part of the dry channel, a proposal with no causal relationship to either fish or wetlands. This was a dramatic instance of the regulatory process gone awry.

In another project, a major fill of tidal wetlands was authorized. The applicant and reviewing agencies had prepared a very clear and detailed account of the relationship of the proposed future mitigation to the values of the wetlands lost. Late in the permit review process the applicant was requested to increase the mitigation acreage and to change substantially the nature of the proposed wetland enhancement. There was no reference to the project record to provide guidance as to the basis for the changes. Hence the changes had the appearance of arbitrary alterations not founded on site-specific analysis. The required mitigation was mandated to be completed on a very short time schedule, but this unrealistic permit condition was ignored by both the permittee and the agencies after the permit was signed. Such decisions remove any incentive for applicants to develop carefully reasoned mitigation plans and undermine the credibility of permit conditions.

Conservationists have no legal standing to challenge the way in which permit conditions overseen by the Corps of Engineers are implemented. Like the setting of Corps wetland regulatory boundaries, the review of mitigation is not a process subject to public input or notice; rather, it is viewed by the Corps as a matter between the agency and the permittee.

The nonexistent or inadequate agency monitoring of onsite construction, once a wetland permit has been issued, long has been a problem. Many replacement wetlands have ended up smaller than they were proposed to be. In one large project near New York City, the fill and commercial development were expanded onto twenty acres of wetlands beyond what the permit actually authorized. Those twenty acres had been slated as part of the mitigation lands when the permit was approved. No agency awareness of this situation was evidenced, even when several years later the Corps issued another permit to the same applicant to fill nearly a hundred more acres of adjacent tidal wetlands. In another project, near Philadelphia, the applicant never completed the initially promised mitigation, but substituted a series of designs for reduced mitigation instead, after the fill permits were approved. Subsequently, a second applicant was authorized to perform mitigation for an unrelated fill on the same site where the first applicant had not completed the promised work. In neither of these cases was the public notified of the substantial changes made in the proposed and initially "required" mitigation. In Florida, too, promised mitigations have often been ignored after the wetlands were filled, despite conditions in the permits (Lewis 1992). Thus it can be difficult to ascertain what is really happening on the ground pursuant to permits that ostensibly require mitigation without actually visiting each project site or comparing clear, as-built drawings and photographs with detailed plans for the mitigation.

There is no question that severely disrupted wetlands can be restored on many sites that formerly supported wetland ecosystems. New wetlands can be created from dry land, given efforts made in good faith by knowledgeable people. Wetlands restored or created by the author more than a decade ago, as well as those built by many others, are functioning as intended. Yet the few available statistics on compliance with mitigation conditions in permits are dismal (e.g., Kantor and Charette 1986; Redmond 1992), causing some conservationists and agency representatives to oppose wetland replacement altogether (Turner 1988).

Technical problems in fact have marked many wetland creation efforts (Kline 1991, Jackson 1990). To build a new wetland may itself constitute a major construction project undertaken reluctantly by a permittee. Design and construction both must be carefully performed if replacement areas are to function successfully as wetland ecosystems appropriate in a given locale. Design choices are many, and trade-offs among objectives should be made explicitly in each case. If the planned or actual grading is incorrect and the water level control is off vertically, functional wetlands at the new land-water interface will not be achieved. Nonnative ground covers may choke out more desirable species of native plants installed at great expense. Wetland designers should closely oversee the on-site realization of their plans. When they do, properly designed new or restored wetlands usually are successful. When new wetland sites have been prepared correctly, the wetland plant communities quickly become established and then undergo slower processes of natural succession. Most new wetlands are intended to be one-time construction efforts that do not require subsequent intervention. If the new ecosystem fails several years later, the permittee generally has no remaining liability that would require trying again.

Where the goal is the maximal achievement of selected wetland values, the wetlands created or restored to compensate for unavoidable fill may or may not closely resemble the wetlands actually lost, at least in the near term. However the new wetlands are designed, their rationale and objectives should be specified. There is no universally valid formula for replacing wetlands. The technical information available for wetland creation is relatively limited, and the existing values of wetlands proposed for loss or enhancement may be underrated, particularly in urban areas (Kraus 1991). Agency and consultant recommendations for wetland plantings sometimes include nonnative plants that are not necessary in wild ecosystems and should not be encouraged there. Plans should be developed based on defined objectives that are appropriate to site conditions and emulate features of nearby existing wetlands. Natural changes should be anticipated in response to the inevitable plant succession on the disturbed site and to the unpredictable series of wet and dry seasons that accompanies any wetland's creation or restoration.

Increasingly, conditions for fill permits are requiring that wetland compensation be monitored and reported to the regulatory agency. This has begun to shift some of the burden of permit follow-up onto permittees. To the extent that designers establish clear criteria for evaluating success and failure, follow their work through the construction process and thereafter for a period of years, and make their findings generally available to the public, the technical craft of mitigation design will be improved. As in other endeavors, reports of failures are often more instructive than reports of successes.

Conclusions

Urban wetlands are scarce and valuable to the public. Combined with waterbodies, they can dramatically enhance the economic value of adjacent lands. Other wetlands, just as valuable for many functions, are not appreciated, and even are discounted by the marketplace. Many wetlands are privately owned, and many currently are in damaged condition. Some may not even be recognized as regulated wetlands.

The complexity of wetland regulation and enhancement has given rise to a need for technical experts familiar with wetland biological, soil, and hydrologic characteristics and with the various regulatory frameworks. The sponsors of construction projects typically must engage competent specialists to identify wetland characteristics and values on land parcels, evaluate potential impacts, work with project designers to avoid wetlands wherever possible, minimize impacts where fill is necessary, compensate for wetland values lost, and monitor the success of attempted mitigation. Project opponents retain experts to challenge developers' assertions, and regulators must decide what is acceptable in specific permit applications. That wetlands add cost and delay to construction projects and are best left alone is now axiomatic wisdom for developers, most of whom avoid wetlands wherever possible. The regulatory requirements for mitigation have motivated research in wetland creation by those who cannot avoid wetland destruction.

The state of knowledge regarding wetlands has been evolving rapidly over the past twenty years. New professional societies, journals, newsletters, textbooks, government agency reports, and voluminous symposium proceedings attest to the technical resources now being devoted to efforts in wetland analysis and enhancement. Wetland regulation in some form appears to be permanent at various levels of government across much of the nation, although the regulatory process is affected by political decisions. For example, after President Bush in 1988 campaigned on a promise of "no net loss," his administration nonetheless spent much more effort dismantling than strengthening federal wetland protection (Goldman-Carter 1992, Dreher 1992). The prospect for more effective wetland replacement is bright, as long as the many professionals involved report their outcomes so that others can learn what works and what should be avoided.

The political decisions that must be made if wetland protection is to be rationalized are several. First, a workable definition of wetlands must be established credibly by federal regulators and maintained for a long enough period so that it can be applied efficiently by federal agencies, by states, by consultants, and eventually by the public. Scientific ("where is the wetland?") and policy ("will the fill be allowed?") considerations should be kept distinct. Second, clear standards are needed for analyzing site alternatives and determining the need for projects affecting wetlands. Third, fair, consistent, and vigorous enforcement against wetland violations is essential.

Technical challenges will always remain regarding the cost-effective creation, restoration, and enhancement of functional wetlands. Increasingly, attention will have to turn to ascertaining how existing wetlands—particularly urban wetlands—can be used for multiple human purposes while maintaining their viability as ecosystems. Most urban wetlands are no longer in pristine condition, and some have been seriously degraded so that they provide values (including aesthetic values) well below their potential. Where urban land values are very high and the functional values of existing wetlands low, it may prove reasonable to allow the filling of some wetlands to provide funds for upgrading the remaining ones, given the probable scarcity of public funds for wetland enhancement in the foreseeable future. On the other hand, if the future acreage of urban and suburban wetlands is to be larger than the existing acreage, new development will have to be responsible financially for most of that increase.

Wetlands are now rare in most cities. The quality of the urban environment no doubt would improve if many remaining urban wetlands were upgraded and if additional wetlands were created to replace some of those lost in years past in every metropolitan area. New wetlands can be established as ornamental landscape elements in commercial, industrial, and residential building projects. They can provide attractive settings for human activity, where wildlife can dwell under the gaze of urbanites, where waters can be purified, and where flowering plants can bloom with the march of the seasons, in sharp contrast to the typical hard surfaces of the built environment. Increasingly, the skillful creation of urban waterways and wetlands has been found by developers to be economically profitable.

In short, the contributions of wetlands to the quality of urban life are far in excess of the proportion of the landscape that they occupy. But at present few new wetlands are being constructed except as required compensation for permits to fill other wetlands. Landowners understandably are reluctant to cause or allow new wetlands to develop on their land because of the potential regulation associated with wetlands. Some jurisdictions regulate the wetlands that can form in stormwater detention facilities; others do not. The present regulatory framework thus serves both directly and indirectly to discourage wetland creation as part of storm-water runoff controls, where new wetlands could accomplish substantial benefits for water quality and the landscape.

The American experience over the past two hundred years proves that public policies can have enormous success in achieving wetland destruction. Those policies included the outright grant of public wetlands to private owners for the purposes of urban development and agricultural conversion, technical assistance for drainage efforts, crop subsidies, public works such as roads and landfills and navigational dredging without controls on the placement of spoils, and tax incentives for replacing wetlands with dry land. Now wetlands are scarce, especially in urban areas. Whether future public policies can successfully reverse the trend—not just by slowing the losses but by actually restoring degraded wetlands and bringing forth new wetlands that achieve multiple functions—for the benefit of our increasingly urbanized society remains to be seen.

Acknowledgments

I would like to thank my longtime colleague, ecologist Stephen P. Kunz, for helpful comments on a draft of this paper.

References

Aurelia, M. 1988. Protecting freshwater wetlands and riparian habitats in southwestern Connecticut. In *Proceedings of the National Wetland Symposium,* ed. J. A. Kusler et al. Oakland, CA, 26–29 June 1988. Berne, NY: Association of Wetland Managers.

Bedford, B. L., M. Brinson, R. Sharitz, A. van der Valk, and J. Zedler. 1992. Evaluation of the proposed revisions to the 1989 "Federal Manual for Identifying and Delineating Jurisdictional Wetlands." Report of the Ecological Society of America's Ad Hoc Committee on Wetlands Delineation. *Ecological Society of America Bulletin* 73(1): 14–23.

Conservation Foundation. 1988. *Protecting America's wetlands: An action agenda.* The final report of the national wetlands policy forum. Washington: Conservation Foundation.

Dahl, T. E. 1990. *Wetland losses in the United States 1780s to 1980s.* Washington: U.S. Fish and Wildlife Service.

Davis, C. 1991. Making no assumptions. *National Wetlands Newsletter* 13(2): 6–7.

Doyle, A. 1986. The Charles River watershed: a dual approach to flood plain management. In *Proceedings of the National Wetland Symposium,* ed. J. A. Kusler and P. Riexinger. Portland ME, 17–20 June 1985. Chester, VT: Association of State Wetland Managers.

Dreher, R. G. 1992. EPA recants role in federal oversight. *National Wetlands Newsletter* 14(5): 10–11.

EL [Environmental Laboratory]. 1987. *Corps of Engineers wetlands delineation manual.* Technical report Y-87-1. Vicksburg, MS: U.S. Army Waterways Experiment Station.

FICWD [Federal Interagency Committee for Wetland Delineation]. 1989. *Federal manual for identifying and delineating jurisdictional wetlands*. U.S. Army Corps of Engineers, U.S. Environmental Protection Agency, U.S. Fish and Wildlife Service, and U.S. Department of Agriculture, Soil Conservation Service. Washington.

Frayer, W. E., T. J. Monahan, D. C. Bowdern, and F. A. Graybill. 1983. *Status and trends of wetlands and deepwater habitats in the conterminous United States, 1950s to 1970s*. Fort Collins: Colorado State University, Department of Forest and Wood Sciences.

Frederick, D. O. 1991. Nature in the hands of accountants. *National Wetlands Newsletter* 13(4): 4–7.

Goldman-Carter, J. L. 1992. The unraveling of no net loss. *National Wetlands Newsletter* 14(5): 12–14.

Jackson, J. M. 1990. *The status of nontidal, freshwater wetlands creation, restoration, and enhancement in the United States; mitigation effectiveness: recap of the literature*. Philadelphia: U.S. Environmental Protection Agency, Region III.

Kantor, R. A., and D. J. Charette. 1986. Wetlands mitigation in New Jersey's coastal management program. *National Wetlands Newsletter* 8(5): 14–15.

Kline, N. 1991. *Palustrine wetland creation mitigation effectiveness*. Philadelphia: U.S. Environmental Protection Agency, Region III.

Kraus, M. L. 1991. The unsung virtues of an urban wetland. *National Wetlands Newsletter* 13(1): 8–9.

Kusler, J. A. 1983. *Our national wetland heritage, a protection guidebook*. Washington: Environmental Law Institute.

Kusler, J. A., S. Daly, and G. Brooks. 1988. *Proceedings of the National Wetland Symposium: Urban wetlands*. Berne, NY: Association of Wetland Managers, Inc.

Kusler, J. A., and E. J. Myers. 1990. Takings: Is the Claims Court all wet? *National Wetlands Newsletter* 12(6): 6–7.

Lant, C. 1993. Making better use of the Farm Bill. *National Wetlands Newsletter* 15(1): 11, 14.

Lewis, R. R. 1992. Why Florida needs mitigation banking. *National Wetlands Newsletter* 14(1): 7.

Nash, R. 1978. Who loves a swamp? In *Strategies for protection and management of floodplain wetlands and other riparian ecosystems*. Proceedings of the symposium, 11–13 December 1978, Callaway Gardens, Georgia. Washington: USDA Forest Service.

NTCHS [National Technical Committee for Hydric Soils]. 1990. *Hydric soils of the United States*. Washington: U.S. Department of Agriculture, Soil Conservation Service.

OTA [Office of Technology Assessment]. 1984. Wetlands: Their use and regulation. U.S. Congress. OTA–O–206. Washington.

PADER [Pennsylvania Department of Environmental Resources]. 1990. *Wetlands protection: A handbook for local officials*. Washington: Environmental Law Institute.

Redmond, A. 1992. How successful is mitigation? *National Wetlands Newsletter* 14(1): 2–6.

Reed, P. B., Jr. 1988. *National list of plant species that occur in wetlands: 1988 national summary*. U.S. Fish and Wildlife Service, Wetland Ecology Group. Biological report 88(24). Washington.

Salvesen, D. 1990. *Wetlands: Mitigating and regulating development impacts*. Washington: Urban Land Institute.

Schmid, J. A. 1990. *Checklist and synonymy of New Jersey higher plants with special reference to their rarity and wetland indicator status.* Media, PA: Schmid & Co.

Thomas, L. 1990. Farmed wetlands: A balancing act. *National Wetlands Newsletter* 12(6): 3, 9.

Tiner, R. W., Jr. 1984. *Wetlands of the United States: Current status and recent trends.* Washington: U.S. Fish and Wildlife Service, National Wetlands Inventory.

Tourbier, J., and R. Westmacott. 1992. *Lakes and ponds.* 2d ed. Washington, DC: Urban Land Institute.

Turner, T. 1988. The myth of mitigation. *Sierra* 73(1): 31–33.

USFWS [U.S. Fish & Wildlife Service]. 1990. *Wetlands: Meeting the President's challenge.* Washington: U.S. Department of the Interior.

USNRC [U.S. National Research Council]; Committee on Restoration of Aquatic Ecosystems; Commission on Geosciences, Environment, and Resources; National Academy of Sciences. 1992. *Restoration of aquatic ecosystems: science, technology, and public policy.* Washington: National Academy Press.

Want, W. 1989. *Law of wetlands regulation.* New York: Clark Boardman.

The Deep Significance of Urban Trees and Forests

John F. Dwyer, Herbert W. Schroeder, and Paul H. Gobster

Trees and forests play a significant role in the urban environment and have many important meanings to urban residents. However, we find that the effort of many urban forestry programs to expand or sustain trees and forests is justified in terms of a few fairly simple dimensions of their significance to urbanites, such as beauty, shade, cooling, or their contribution to global gas balances. Programs based on this narrow spectrum of benefits may not fully meet the needs of urbanites or gain their support. We recommend a broader perspective, one that takes into consideration the deep psychological ties between people and urban trees and forests. In this paper we outline some of the major ties we have found, and suggest their implications for the management of urban forestry programs.

For the past decade, we and a number of colleagues have been studying the role that trees and forests play in people's preferences in and perceptions and uses of urban environments. These efforts have shown the importance of trees and forests in many kinds of urban settings, including streets, housing developments, parks, forest preserves, arboretums, botanic gardens, conservatories, and trail corridors (Schroeder 1989). We have found consistent patterns in people's responses and have built statistical models to help managers predict how urbanites will respond to options in the management of urban forest resources.

Much of our initial work focused on predicting public preferences for changes in urban forest management, for example, planting new trees along a street, increasing the density of the tree cover in a park, or altering the landscape along a bicycle trail. As we conducted this work, however, it became increasingly clear to us that the values of trees and forests in urban areas involved more than simple pleasure in the attractive environments they provide.

There are deep emotional ties between people and trees that are not conveyed by a high correlation between "tree size" and "preference" in the equations we have developed to predict perceptions of urban forest environments. Likewise, the strong ties between people and trees cannot be explained by increased property values, reductions in air pollutants, and moderation in temperature. The psychological ties between people and trees defy easy quantification, yet few would deny their existence or their profound implications for urban forest management. We come across these ties in many aspects of our everyday experience with trees and in our daily contact with individuals. Together these ties help us understand the deeper meaning of people-tree relationships. Among the many indications of public affection for trees are:

—frequent calls to our office (USDA Forest Service) for help in saving trees threatened by insects and disease;
—yellow ribbons tied to trees to protect their removal for roadways or other purposes;
—trees often planted as "living memorials" to loved ones;
—references to trees in the names of communities and subdivisions (e.g., Tall Oaks Condominiums, Village of Elmhurst);
—efforts by adults to provide urban children with forest experiences in urban nature centers, parks, arboretums, botanical gardens, or zoos.

Strong bonds between people and the natural environment have probably been discussed most widely with respect to wilderness and rare and endangered plants and animals. The support for the preservation of wild places and the species of plants and animals that live there is often very strong—and sometimes especially strong among urban people who have never seen these places and never expect to. Some of this support has been attributed to the attraction that many urbanites feel for a natural landscape that is distinctly different from the highly developed environment in which they live. Urban trees and forests provide contrast and relief from the highly built-up city environment, but they often lack the aura of "pureness" associated with the remote natural areas of this and other continents. However, the close bonds between people and their urban trees and forests may be enhanced by almost daily contact during all seasons of the year and by the distinct contrast between trees and the built-up environment. The role of urban nature as a relief from the developed environment is illustrated in one person's account of an experience in Lincoln Park on the Chicago lakefront. (This and all subsequent quotes in this paper were taken from written responses to questions concerning individuals' experiences involving trees and forests.) "While bicycling in Lincoln Park, Chicago in the fall, I came upon a scene of colored trees next to a lagoon [pond]. The view was totally natural—water, trees, and bushes—no buildings. It was very beautiful—at least to a city person who sees brick, asphalt, and concrete most of the time."

Efforts at preservation also appear to be motivated by the feeling that trees and forests are "threatened" or "vulnerable" in the urban environment. In other instances trees are valued for their enduring quality, as in the following account by a University of Wisconsin student: "I focused on a huge tree that was situated about fifty yards ahead of me. . . . As I approached the tree, I noticed it was a Bur Oak of an unbelievable size. . . . I became an insignificant figure . . . I wondered how many years the tree must have been there. . . . As I walked away, I took one last glance of a 'King' that has never given in to man and other vegetation."

Research on the Significance of Trees and Forests

To explore in depth the strong attractions between people and trees, we have conducted studies in which individuals were asked to describe significant environments and experiences and to explain the meanings they associate with those environments (Schroeder 1988a; Chenoweth and Gobster 1990; Gobster and Chenoweth 1989, 1990). Much of this research has been conducted at the Morton Arboretum, a nature sanctuary that lies at an intermediate position along an urban-wilderness continuum. Its fifteen hundred acres in the western suburbs of Chicago range from open mowed grassy areas and formal plantings to prairie and some quite "natural" woodlands—all readily accessible by road. Our studies suggest that this area can provide many of the experiences people often associate with wilderness.

The results of our research revealed some very profound emotional ties between people and forests. An example is the following description of an experience at the Morton Arboretum (Chenoweth and Gobster 1990):

A good friend had recently lost a loved one and was feeling extremely depressed. It was about 4:00 P.M. on a warm and sunny Autumn day. Being familiar with the Morton Arboretum and with its beauty at this time of the year, I felt that a drive through the Arb could be both pleasant and therapeutic. She agreed to go. As we drove into the Arb she remarked about the changing colors of several individual trees. It was almost peak fall color. While riding, we talked freely of our feelings and her present situation. As we approached the Forest area, I chose a road with no other cars or people in sight. We were able to drive slowly and soon came to the densest part of the forest where the sugar maples had turned brilliant colors of yellow and orange. Mingled in with the maples were tall green spruces; the Virginia creeper with its fall red coloring dappled the other colors. It was as if, suddenly, we were inside a large cathedral with stained-glass windows. The feeling was magnificent and awe-inspiring. Almost automatically my car came to a stop. All conversation came to a stop. The "peak" aesthetic experience occurred as the presence of a Supreme Being seemed to engulf us. The beauty of the environment and the solitude of the forest made us become "one." We were quiet and motionless for several minutes. A few tears rolled down the cheek of my friend. Quietly, she said, "Thank you, I feel better—I can face anything now." It was a profound experience for both of us.

In the rest of the paper we present a number of themes concerning the ties between people and trees and forests that we have discovered in our research, from our observations, and in the findings of others. These themes help us explain and interpret the results of our research on people-environment interactions and may provide useful insights for foresters, arborists, landscape architects, and others who manage urban forest resources.

Sensory Dimensions of Trees

The contribution of trees and forests to the beauty of the urban environment is well documented (Schroeder 1989), but their influence on urbanites goes deeper than visual aesthetics. Trees and vegetation can have a strong relaxing effect on people. Four-fifths of the respondents in a study of Morton Arboretum users described their favorite settings as "serene," "peaceful," and "restful" (Schroeder 1988a). Morton Arboretum visitors have described the calming effect of the arboretum in terms such as the following:

> A pleasant tranquil place to be—rather like a living piece of artwork in its impact. It appears orderly, in balance, and inviting.

> The "scape" across the valley and up the distant rise in autumn brings such a feeling of serenity. It makes me feel that "God's in His Heaven" and "all's right with the world." Each shade of leaf coloring is Art in itself. I could sit here forever and let the world go by.

> A feeling of quiet, peace, and order arises within me.

Roger Ulrich (1981) and his associates have actually measured the relaxation effects associated with views of trees and other plant life. They found that individuals who viewed urban scenes with vegetation had slower heartbeats, lower blood pressure, and more relaxed brain wave patterns than individuals who viewed scenes without vegetation. Similarly, Ulrich (1984) reported that hospital patients recovering from surgery who had a view of a grove of trees through their windows required fewer strong pain relievers, experienced fewer complications, and left the hospital sooner than similar patients who only had a view of a brick wall.

The individuals in Ulrich's studies responded to the visual qualities of vegetation viewed in pictures or through a window. What about responses to the many other sensory dimensions of the forest environment such as sounds, odors, shelter, and lighting? What about the sound of wind rustling through leaves or branches? This sound is similar to "white noise," which has been used in hospital wards to mask disturbing sounds and help heart attack victims to relax. Recordings of wind and rain in the trees have been marketed as aids for meditation and relaxation (Schroeder 1991). What about the sounds of birds and insects in the trees? What about the odors associated with trees and forests, including the fragrance of flowers and understory plants, the distinct essence of the humus layer in a pine forest after a rain, or the aroma of wet leaves in the fall? Research has shown that compared with visual stimuli, smells tend to trigger responses that are more emotional and cognitive, and that they tend to be remembered much more vividly than the visual characteristics of places (Porteous 1985). What about the feeling of being sheltered or protected from the sun, wind, and rain by a natural "roof over your head"—feelings that go well beyond simple physical sensations to reach innermost feelings? What about the ever-changing patterns of color, light, and texture created as light flickers through leaves and branches—alternately revealing and hiding other components of the urban environment (Appleyard 1980)?

Trees and forests reach out to the urbanite and convey serenity and beauty in a number of sensory dimensions, and often surround the individual with nature in an environment where natural things are few. This characteristic is reflected in a line from a popular John Denver song of a few years ago: "You fill up my senses like a night in the forest." Trees and forests, and especially large trees or groups of trees, touch our lives in so many ways that it is difficult to describe them. Synergism among these attributes often complicates our attempts to understand the ties between people and trees and forests. The task is similar to describing what attracts us to a loved one, a home, or a profession. We can come up with endless lists of reasons but are seldom satisfied that we have included all the dimensions, or that they collectively come anywhere near capturing the totality of what we are evaluating.

The following statements by a Morton Arboretum visitor illustrate the profound influence that trees can have in a residential environment:

> Elm trees were part of my life. We all loved the elms for their beauty, shade, and protection from rain, but never fully appreciated them until they started to die. As each one died in my neighborhood, it left an irreparable scar and the place began to look old, worn, and crowded. When the enormous elm in our front yard showed signs of disease, we moved—after 20 years of residence. That old neighborhood became somewhat of a slum. But in recent years

people have been renovating the homes and it is again considered a "nice" place to live. . . . [I]ncidently the replacement trees for the elms are now mature. . . . When my husband and I bought our first and present home (he is from Wisconsin) we insisted on an older home with mature trees. We settled on the older area of Glen Ellyn. We have planted 12 additional trees.

Each sensory aspect of the urban forest changes over time—in the short run with the seasons, and in the long run with tree growth, maturation, and death. This evolution is a key element in the psychological ties between people and trees. Comments of Morton Arboretum visitors about their favorite settings convey the importance of the dynamic qualities of the urban forest:

It is always changing naturally as well as by [the acts of] man.

. . . an always changing and growing place.

Every two weeks different flowers are blooming and the plants are all a lot taller. It is a source of endless fascination to me.

Reminds me of the first breath of spring after a long winter and the reawakening of nature.

The changing colors in the fall foliage warns our hearts that wintry winds will soon be upon us and we must enjoy the present as long as it lasts.

Seeing the rich growth from the decay of other vegetation reminds me of the cycles of life and death. Death is a form or part of the life cycle.

Symbolic Values of Trees

Apart from the sensory experiences they provide, trees are often valued as carriers of symbolic meanings. There are many examples of trees considered as representing human qualities and trees regarded as religious symbols.

Trees as Symbols of People

Donald Appleyard (1980) observed several parallels between our images of people and of trees. The sheltering quality of trees suggests a parental nature. Old trees look wise, and young saplings are fresh and growing. We feel sad when a tree looks sick. We speak of a tree's branches as "limbs," as we do of our own arms and legs. Some leaves are characterized as "reaching out like fingers" or having a *palmate* shape (after the Latin word for hand), while in the fall others "curl up like a fist."

Cartoons depicting trees often show the branches as arms and the trunks with faces. When a tree has been damaged, we speak of "wounds" and "healing," and may enlist the help of a tree "doctor" or "surgeon." We speak of our own "roots" and our "family tree."

Appleyard (1980) further observed that trees are seen as being innocent, fragile, and endangered, and as demanding our concern and pity. They are one of the most loved elements of the urban environment. It is no accident that advocates and supporters of urban trees are often referred to as "tree huggers."

Several writers have made the analogy between the individual, cultural, and social characteristics of trees and those of people. Single trees planted in parks and yards grow quite differently than trees growing together in a forest setting. Like people, individual trees have their own unique appearance, personality, and idio-syncracies. The great diversity of tree species and varieties mirrors our own ethnic and cultural diversity as expressed in appearance, customs, and traditions. And when trees grow together in forests, we tend to focus on the qualities of assets of the whole group, much in the same way that we look at the qualities or assets of a community, business, or other group of people. In his book *The Tree,* author John Fowles (1989) parallels the "sociology of trees" with the human condition and the value we place on independence while recognizing the necessity to belong and cooperate: "We think we feel nearest to a tree's 'essence' (or that of its species) when it chances to stand like us, in isolation, but evolution did not intend trees to grow singly. Far more than ourselves, they are social creatures, and no more natural as isolated specimens than man is as a marooned sailor or hermit."

Perhaps our love of trees and their similarities to people are responsible for our efforts to plant trees as "living memorials" for loved ones. Trees are common in cemeteries, and many trees and forests have been planted as memorials to fallen soldiers and others. Our office has received several calls from people requesting information about planting commemorative trees. One woman wanted to know how to propagate a buckeye tree that stood near her grandfather's home. The grandfather had passed away and now the tree appeared to be dying. Each family member wanted two seedlings from "grandpa's tree" to remember him by. Others have called to find out more about a certain casket company that provides a certificate to the family of the deceased indicating that a tree has been planted in a National Forest in the person's name.

Trees as Religious Symbols

Trees have been used by many cultures to symbolize health, wisdom, and enlight-enment (Schroeder 1988b). Holmes Rolston (1988) describes a gnarled, windblown tree as representing endurance and strength and as symbolizing life pushing on before the winds that blast it. Herb Schroeder (1988b) provides illustrations of a number of religious and cultural traditions where trees stand as a symbolic link between the human and divine, and are the means by which humans come into contact with their deepest spiritual values. This spirituality is exemplified by the two trees in the Garden of Eden (the Tree of Life and the Tree of Knowledge) in the Hebrew creation story. The Christian cross is sometimes identified with the Tree of Life; and in the Book of Revelation, the Tree of Life is found growing in the New Jerusalem. Hindu symbolism represents the awakening of divine consciousness as a serpent ascending a tree, and Buddha is reported to have achieved enlightenment while sitting under the "wisdom tree."

Rolston (1988) and Schroeder (1991) refer to the forest as a religious resource and compare forests to places of worship such as cathedrals. The spiritual-religious values of wilderness have long been noted. A question remains concerning the extent to which urban trees and forests are associated with religious feeling. Several studies suggest that urban trees and forests can contribute to experiences that are religious in nature (Chenoweth and Gobster 1990; Gobster and Chenoweth 1989, 1990; and Schroeder 1988a, 1988b). The following comments from Morton Arboretum visitors support such a view:

> Being deep in the woods is a place where your spirit can fly free, without interruption, bringing you closer to God.

> Each time I look up to one of those tall trees at the Arboretum, I am reminded of Joyce Kilmer's poem, "Trees." I too look to God and pray.

> . . . trees towering over the road like a cathedral . . .

> There is at once strength, and form in these native trees that create a potent force and magic in the area.

Human Roots in the Forest/Savanna

People's responses to trees and forests are so strong and consistent that some researchers have even suggested that human beings have evolved instinctive preferences for certain types of treed environments. Most humans appear to prefer groves of widely scattered trees, open at eye level, with an overhead canopy and a uniformly textured ground cover (Kaplan 1984). It has been said that this environment may be attractive because it resembles African savannas where the human species evolved (Balling and Falk 1982). In that sense it may represent our most primal image of "home" (Schroeder 1991). Holmes Rolston (1988) also considers it significant that humans evolved in forests and savannas and "love the forest for what it aboriginally is." Some say that people go to the forest "to get away from it all"; Rolston suggests instead that they "go there to get back to it all."

Trees are one of the last representatives of nature in the city, providing a constant reminder of the natural world beyond the city as well as of humanity's distant past (Appleyard 1980). Some people tell us that when they are in an urban natural area, they fantasize about traveling back in time to earlier experiences in their lives or to a time when there were no cities on our continent. For them the urban forest offers an opportunity to escape from daily routine and urban stress, a sentiment expressed by Morton Arboretum visitors:

> . . . a place of beauty, peace, quiet excitement and refuge from the noise, turmoil, pollution and unpleasantness of traffic and crowded work and living conditions.

> A forest represents to me cool, calm, a place to regain composure.

> I think of this as a place to contemplate, to stop and use my senses, to remove myself from today's schedule of events.

> I felt I was somewhere quite far away from the bustle and noise of people and cars.

> The Arboretum has been a resource and refuge for me through what were often some difficult times.

> All problems and cares disappear for the time being.

Students from the University of Wisconsin have also implied an element of escape as part of their experiences with urban trees and forests:

> While walking through Tenney park on my way to the library I became more and more absorbed in the fresh snow and beautiful landscape until I felt part of the whole scene. I stopped, stared at a grove of trees and felt a wonderful calming sensation. It was as though time had momentarily stopped and there I was—part of the whole park scene.

> I looked up into the miraculously blue sky through the still unleafed branches of the oak tree. The twigs and branches in front of the blue background formed a very delicate pattern. I never looked at a tree like that before. There was beauty in it just for the sake of it. I looked at it, I could not get enough of it, and for a short time I forgot the surroundings—the activities that were going on around me.

Fear of the Forest

It is important to recognize that the images of trees and forests in our country's past have not always been positive. They include images of the "howling wilderness" full of savage beasts and other dangers. In the early years of European settlement in this country, the forest was often a barrier to cultivation and a hiding place for enemies. Fears of the forest persist to this day. Some have their roots in the past, but others are a function of more recent concerns.

In our research we have encountered examples of deep-seated fears of trees and forests. Although these fears are not necessarily widespread, many urbanites are afraid of going into a forest area. City dwellers may fear being attacked by criminals who hide in urban vegetation, becoming lost while in a forest, or contracting Lyme disease. These fears may cause people to limit the density of tree planting around homes and to avoid heavily forested portions of the city. We suspect that fear of getting lost may cause many urbanites to walk through the forest on paved bicycle trails and to prefer forest scenes with large open areas or buildings in them. For some time we have observed a wider variety of users on urban bicycle trails than in the surrounding forests. Other fears associated with trees include concern that animals and insects will invade the home or that trees or parts of them will fall on a home or car.

Urban African Americans are not generally found to be heavy users of forest areas, and studies show they are more likely than urban whites to prefer developed recreation areas over more natural ones (Dwyer and Hutchison 1990). Our research with African Americans in the Chicago area suggests the persistence of negative images of forests in rural areas of the South, where many have their roots. The concerns range from the presence of insects, snakes, and other creatures to threats to life and limb. We have found particularly strong fears of forest environments among female children of urban African Americans (Metro et al. 1981).

These fears may in part reflect a lack of familiarity with forests, and their origins probably relate to the environments in which these people were raised. For example, in one study individuals who grew up in suburban areas tended to feel most comfortable in natural settings, while those who grew up in cities tended to feel most comfortable in developed settings (Schroeder 1983).

Tree Planting as an Activity

Human emotional ties to trees express themselves in many ways. One manifestation of particular interest to urban foresters is the activity of planting trees. Tree planting is a popular activity among urbanites; the level of interest has apparently been increasing in recent years and was particularly strong during the recent Earth Day/Arbor Day celebrations. The popularity of and commitment to tree planting suggest that it has benefits that go beyond the benefits expected from the resulting trees (i.e., tree planting is a good thing to do even if the trees don't survive!). Some possible explanations for the strong interest in tree planting include:

Tree planting as a demonstration of commitment to the future. It has often been said that one plants a tree not for oneself but for future generations. Perhaps this explains why we encountered so many multigenerational groups at tree plantings this past year. In our highly mobile society, tree planting is often taken as a symbol of permanence or "putting down roots." Appleyard (1980) characterized trees as "anchors of stability in the urban scene."

Tree planting as a contribution to the landscape over time. Few activities that individuals can undertake on their own land have the potential for as large an impact as tree planting does. A small seedling can grow to a huge oak that dominates the local landscape for a very long period of time. People may not see that happen in their lifetimes, but they may get some hint of it as the tree grows, and take pleasure in the belief that their action will have influence long after they have departed from the scene. The tree may also provide satisfaction in the shorter run, because the types of trees and other plants are often the major distinguishing variable among homes in a new subdivision.

Tree planting as a means of improving the environment. With increasing recognition of numerous global environmental problems, individuals can often feel helpless in the fight for a better environment. Tree planting is one thing they can accomplish on their own and by doing so feel that they have helped solve some of the problems. Discussion of the possible role of trees in reducing urban heat islands, global warming, noise pollution, air pollution, erosion, fuel consumption, and other problems most likely reinforces this motive for tree planting. Tree planting puts into practice the motto "Think globally, act locally," which has become popular with citizen groups. In Chicago, citizen groups have been actively involved in tree planting under the Open Lands Project's "Neighborwoods" program as well as Mayor Daley's "Greenstreets" program.

Unfortunately, the deeply held values that motivate people to plant trees often do not find expression in a desire to care for the trees on a regular basis. Consequently we often see much effort given to urban tree planting, only to have some of the trees deteriorate or perish from lack of watering and basic maintenance. Perhaps more attention should be given to providing information on how tree care practices are tied to tree preservation. This might include involving urbanites in the care of public trees.

We also see urban tree-planting efforts that do not capitalize on many of the benefits that could be generated by urban trees—for example, shading and reduced heating and cooling costs (Schmid 1975) (see McPherson paper). At times tree planting seems to be just for the sake of tree planting, and the trees are planted too far from the house to shade the roof, windows, or air conditioner. There may be a need for better dissemination of information on how species selection, placement, and care can provide particular benefits. Perhaps the increased involvement of urbanites in neighborhood and community tree-planting efforts will afford additional opportunities for disseminating such information.

Summary

The importance of urban trees and forests is grounded in some very deep feelings and ties that sometimes even have a spiritual quality. These feelings manifest themselves in strong support for tree planting and for the preservation of existing trees and forests. They may also emerge in preferences for or fears of particular urban forest environments. These powerful forces must be taken into account in any effort to expand and enhance urban forest resources.

Perhaps the strong interest in tree planting can motivate urbanites to reforest many of the cities where the forest has been depleted. At the same time we hope urbanites can be provided with information about the systematic professional tree care and maintenance that must accompany tree planting.

The values mentioned in this paper tend to evoke emotional rather than rational arguments for tree preservation. Managers, who are trained to think in terms of rational justifications for the money for tree-planting and maintenance programs, need to see the other side of the coin. Managers and administrators often report that trees and tree care cannot compete well with other functions provided at the municipal level. Businesses that employ tree care professionals face similar challenges concerning the allocation of scarce resources. Perhaps part of the solution to this problem is to focus attention on new information and perspectives concerning the deeply held values that trees represent for people.

To discover what urbanites value, we need to listen more and preach less. Education programs for urbanites are useful, but we managers, planners, and researchers also need more education to understand the basis for people's behaviors, preferences, and fears. Research on perceptions, values, and behaviors concerning urban trees and forests can play a key role in the educational process.

Our studies of the psychological values of trees in urban areas confirm the importance of "everyday nature" as a significant contributor to the health and well-being of the urban population. People should not have to leave the city to find opportunities to refresh themselves in nature. This research shows the unique value of experiences of nature "at the doorstep", of trees and nature in our everyday lives.

Our work does not tell managers such things as how many trees to plant and where to plant them. It does, however, provide support for overall program emphases as well as ideas for "marketing" these programs to the general public and to policy-makers. In many cases it points out that what people value in trees is not so much the tangible benefits such as energy savings or property values as the experiences that trees offer.

At present we lack information on the best design and management of urban forests for providing important benefits such as stress reduction and opportunities for recreation and leisure activities, or for creating particular sights, sounds, or smells. We also need to understand people's fears and concerns about problems associated with trees. With this knowledge we might change the way that urban trees and forests are managed, and also furnish urbanites with information to help them make appropriate choices concerning the management and use of urban forests.

In sum, urban trees are living, breathing organisms with which people feel a strong relationship, and in our planning and management we should not think of them just as air conditioners, providers of shade, and ornaments in the urban system. Failure to recognize the deep significance of trees to urbanites will most likely result in giving less effort to tree planting, care, and protection than the public desires.

References

Appleyard, D. 1980. Urban trees, urban forests: What do they mean? In *Proceedings of the National Urban Forestry Conference.* 13–16 November 1978. Syracuse: State University of New York College of Environmental Science and Forestry.

Balling, J. D., and J. H. Falk. 1982. Development of visual preference for natural environments. *Environment and Behavior* 14:5–38.

Chenoweth, R. E., and P. H. Gobster. 1990. The nature and ecology of aesthetic experiences in the landscape. *Landscape Journal* 9(1): 1–8.

Dwyer, J. F., and R. Hutchinson. 1990. Outdoor recreation participation and preferences by black and white Chicago households. In *Social Science and Natural Resource Recreation Management,* ed. Joanne Vining. Boulder, CO: Westview.

Fowles, J. 1989. *The Tree* (excerpt). In *Trees: A Celebration,* ed. J. Fairchild. New York: Weidenfeld and Nicolson.

Gobster, P. H., and R. E. Chenoweth. 1989. The dimensions of aesthetic preference: A quantitative analysis. *Environmental Management* 29:47–72.

———. 1990. Peak esthetic experiences and the natural environment. *Proceedings of the Twenty-First Annual Conference of the Environmental Design Research Association,* ed. R. I. Selby, K. H. Anthony, J. Choi, and B. Orland. 6–9 April 1990. Champaign-Urbana, IL.

Kaplan, R. 1984. Dominant and variant values in environmental preference. In *Environmental Preference and Landscape Preference,* ed. A. S. Devlin and S. L. Taylor. New London: Connecticut College.

Metro, L. J., J. F. Dwyer, and E. S. Dreschler. 1981. *Forest experiences of fifth-grade Chicago public school students.* Research pap. NC-216. St. Paul, MN: U.S. Department of Agriculture, Forest Service, North Central Forest Experiment Station.

Porteous, J. D. 1985. Smellscape. *Progress in Human Geography* 9(3): 356–78.

Rolston, H. 1988. Values deep in the woods: The hard-to-measure benefits of forest preservation. In *Proceedings of the 1987 Society of American Foresters National Convention,* 18–21 October 1987. Minneapolis.

Schmid, J. A. 1975. *Urban vegetation: A review and Chicago case study.* Research paper no. 161. Chicago: The University of Chicago Department of Geography.

Schroeder, H. W. 1983. Variations in the perception of urban forest recreation sites. *Leisure Sciences* 5(3): 221–30.

———. 1988a. The experience of significant landscapes at Morton Arboretum. In *Proceedings of the 1987 Society of American Foresters National Convention,* 18–21 October 1987. Minneapolis.

———. 1988b. Psychological and cultural effects of forests on people. In *Proceedings of the 1988 Society of American Foresters National Convention,* 16–19 October 1988. Rochester.

———. 1989. Environment, behavior, and design research on urban forests. In *Advance in environment, behavior, and design.* Vol. 2, ed. E. V. Zube and G. T. Moore. New York: Plenum.

———. 1991. The psychological value of trees. *The Public Garden* 5(1): 17–19.

Ulrich, R. S. 1981. Natural versus urban scenes: Some psychophysiological effects. *Environment and Behavior* 13:523–56.

———. 1984. View through a window may influence recovery from surgery." *Science* 224:420–21.

Cooling Urban Heat Islands with Sustainable Landscapes

E. Gregory McPherson

Introduction

The rapid urbanization of U.S. cities during the past fifty years has been associated with a steady increase in downtown temperatures of about 0.1° to 1.1°C (0.25° to 2°F) per decade. Because the demand of cities for electricity increases by about 3 to 4 percent for every increase of one degree Celsius (1.5 to 2 percent per degree Fahrenheit), about 3 to 8 percent of current electric demand for cooling is used just to compensate for this urban heat-island effect (Akbari et al. 1990). Other implications of growing urban heat islands include increases in carbon dioxide emissions from power plants, municipal water demand, concentrations of smog, and human discomfort and disease. Global warming, which may double the rate of urban temperature rise, could accentuate these environmental problems. More-over, the accelerating world trend toward urbanization may expand the local influence of urban heat islands, as megalopolises begin to modify regional climate and airflow (Tyson et al. 1973).

This paper is directed to the policy-makers who are responsible for urban design and its climatological consequences. It summarizes our current knowledge on the structure, energetics, and mitigation of the urban heat island. Special attention is given to physical features of the environment that can be easily manip-ulated, particularly vegetation. Prototypical designs illustrate how concepts of sustainable landscapes and urban climatology can be applied to counteract urban warming in street canyons, parking lots, urban parks, and residential streets. In a previous study (McPherson 1990a), sustainable landscapes were defined as multi-functional, low maintenance, biologically diverse, and expressive of "place." Miti-gation of urban heat-islands by landscapes can contribute to the sustainability of our cities. Because most electric utilities experience peak demands during summer because of air-conditioning loads, this paper addresses mitigation of summertime heat islands, while recognizing that winter heat islands can be beneficial in cities with cool climates.

Urban Heat Islands

Warmer air temperatures in cities compared to air temperatures in surrounding rural areas is the principal diagnostic feature of the urban heat island. Alterations of the urban surface by people result in diverse microclimates whose aggregate effect is reflected by the heat island (Landsberg 1981). Buildings, paving, vegetation, and other physical elements of the urban fabric are the active thermal interfaces between the atmosphere and land surface. Their composition and structure within the urban canopy layer, which extends from the ground to about roof level, largely determine the thermal behavior of different sites within a city (Goward 1981; Oke 1987a). Thus, urban heat islands can be detected at a range of scales, from the microscale of a shopping center parking lot to the mesoscale of an urbanized region.

Structure of Urban Heat Islands

The structure of urban heat-islands has been well documented from climatological studies of cities around the world (see Chandler 1965; Landsberg 1981; Oke 1986). Urban and rural temperature differences are greatest and the spatial and temporal qualities of these anomalies most apparent during clear and calm summertime conditions. The horizontal structure of a hypothetical heat island is characterized by a "cliff" that follows the city's perimeter and is steepest along the windward boundary (Oke 1982). This sharp temperature gradient leads to pulses of cool air flowing into the city at night. Intraurban heat islands and "cool islands" reflect localized effects of differences in building density and surface cover. Temperatures in mid-latitude parks can be 1° to 3°C (1.8° to 5.4°F) cooler than outside, and their influence can extend several hundred meters beyond the park boundary (Chandler 1965; Herrington et al. 1972; Oke 1989). Differences in urban and rural temperatures usually are greatest (3° to 8°C) in early evening near the city core. However, daytime temperatures often are warmest outside the core in a zone with lower buildings and more exposed pavement (Tuller 1973). Winds carry the warmth of the city downwind.

Analysis of temporal differences shows that the intensity of the urban heat-island is greatest at night, primarily due to differences in urban-rural cooling (Oke 1982). Nocturnal urban air-temperature anomalies of 3° to 5°C are typical, as compared with 1°C daytime anomalies (Goward 1981). At sunset, rural areas begin to cool rapidly while urban areas remain warm and then cool at a slower rate. Different urban-rural cooling rates at sunset produce maximum heat-island intensities three to five hours later. At sunrise, urban areas begin to warm relatively slowly, sometimes producing urban "cool islands" during the morning.

Under calm conditions, a rural/urban breeze system develops at night that modifies the heat island's vertical structure by creating an urban heat dome. Downwind heat plumes can extend over rural areas for considerable distances. The vertical extent of air temperature anomalies at night is only two to three times building height, compared with more than 1 kilometer during the day. Increased turbulent mixing of the atmosphere during daytime is primarily responsible for this urban impact on the atmosphere (Duckworth and Sandberg 1954).

Energetics of Urban Heat Islands

Radiation and anthropogenic sources of energy are partitioned into latent, sensible, and stored energy within the urban environment. Flows (or fluxes) of energy for each of these terms are expressed in the energy balance equation for an urban surface as:

$$Q_F + K^* + L^* = Q_F + Q^* = Q_E + Q_H + Q_S$$

where

Q_F is anthropogenic heat release,
K^* is net shortwave radiation (direct, diffuse, and reflected),
L^* is net longwave radiation,
Q^* is the net allwave radiation,
Q_E is latent heat-flux density,
Q_H is sensible heat-flux density, and
Q_S is net storage heat-flux density.

The magnitude of these energy fluxes ranges widely within and among cities depending on factors such as city size, population, latitude, urban morphology, and land cover characteristics. As a general guide for assessing the relative importance of each flux along the urban-rural gradient, Oke (1988a) listed hypothetical flux densities based on existing observations. These data, reproduced in table 1, refer to clear summer conditions at noon for a mid-latitude city with 1 million inhabitants. Unfortunately, typical values for nocturnal conditions, when the urban heat island is most intense are unavailable.

The relative importance of anthropogenic heat is small except during winter in high-latitude cities. At noon during summer, anthropogenic heat-flux densities typically range from twenty to fifty watts per square meter (table 1), depending on population density and energy use per capita (Oke 1988a).

Surprisingly, differences in net all-wave radiation (Q^*) between rural and urban environments are relatively small because of offsetting the effects of urban atmosphere, albedo, and geometry (table 1). On average, smog attenuates urban short-

Table 1   Radiation and Energy Balances for Hypothetical Midlatitude City at Noon During Clear Summer Conditions. All Values in W m^{-2} (after Oke, 1988a).

Available Energy	Rural		Suburban		Urban	
$K\downarrow$	800		776		760	
$K\uparrow$	−160		−116		−106	
$L\downarrow$	350		357		365	
$L\uparrow$	−455		−478		−503	
Q^*	535		539		516	
Q_F	0		15		30	
Total	535		554		546	

Partitioned Energy	Rural	%	Suburban	%	Urban	%
Q_S	80	15	122	22	148	27
Q_H	150	28	216	39	240	44
Q_E	305	57	216	39	158	29
Total	535	100	554	100	546	100

wave radiation by about 10 percent, but a lower urban albedo (15 percent versus a rural albedo of 20 to 25 percent) results in relatively more absorption than in rural surroundings. Longwave radiation is primarily a function of surface temperature and exposure to the sky. Warmer air temperatures above cities result in greater incoming longwave radiation, but this is offset by increased outgoing longwave radiation due to warmer urban surface temperatures. However, it is important to recognize that urban heat-island intensity is greatest in urban canyons, partly because of the effect of tall buildings on radiation exchange. These buildings shade the street and wall surfaces during the day, reducing absorption of shortwave radiation and in some cases creating daytime "cool islands." However, the rate of heat loss to the sky at night is reduced because the sky in the city core is a less accessible sink for longwave heat loss compared with the sky in more open rural or suburban settings. Thus, although net allwave radiation in urban and rural environments is similar when averaged over a day, the geometry of cities results in diurnal differences in flux densities that can have important effects on the timing and intensity of the heat island.

Available energy ($Q_F + Q^*$) is partitioned into three fluxes, Q_S, Q_H, and Q_E. Net storage heat flux (Q_S) depends on the thermal properties and arrangement of elements within the urban canopy layer. The thermal properties of typical urban materials such as asphalt, brick, and concrete are similar to those of rural materials such as bare soils. Higher urban net storage heat-flux densities are attributed to differences in the composition and structure of these materials rather than their thermal properties per se (Goward 1981). For instance, unshaded, vertical building walls in the urban core can absorb larger amounts of radiant energy per unit of ground surface area than flatter rural surfaces can. Much of this energy is returned to the air as sensible heat because the low thermal mass of honeycomb-structured buildings results in relatively little heat storage. Paving materials have high thermal inertias that delay the time of peak surface temperature and reduce temperature ranges compared with those for surfaces of unshaded buildings with similar albedos. For the hypothetical mid-latitude city at noon, net storage heat flux typically accounts for about 15 and 27 percent (80 and 148 watts per square meter) of the available energy in rural and urban areas, respectively (Oke 1988a). Most importantly, daytime heat storage by the urban fabric delays the onset and retards the rate of nocturnal cooling.

Partitioning of the remaining energy into sensible (Q_H) and latent energy (Q_E) fluxes depends largely on available atmospheric, surface, and soil moisture. Moisture availability in urban environments exhibits great diversity in both space and time, whereas spatial diversity is less in rural areas (Oke 1988a). Lakes, irrigated fields, and forests of rural areas transform incoming energy into latent energy through evapotranspiration. Because less energy remains as sensible heat, air temperatures are lower. Hence, at noon in rural areas, 50 to 60 percent of the available energy is consumed as latent heat flux, and about 30 and 15 percent become sensible and subsurface heat fluxes (table 1). Vegetation cover accounts for from 5 to 20 percent of the surfaces in cities, 15 to 50 percent in suburbs, and 75 percent or more in rural areas (Rowntree 1984). Because cities have less vegetation and more impervious surfaces than rural areas, more energy is available to warm the air. Latent heat flux typically consumes only 30 percent of the incoming energy in the irrigated hypothetical city at noon, about half that consumed in the country. Hence, in urban areas, nearly 45 percent of the incoming energy warms the air as sensible heat (table 1).

It should be recognized that in the previous discussion "ideal" conditions were assumed. When conditions are calm and clear, microclimatic differences are enhanced and the relative importance of radiation and heat storage is increased. As windspeed and cloud cover increase, convection becomes relatively more important and net radiation and storage heat fluxes are dampened.

Implications of Urban Heat Islands

Urban temperatures have been increasing in cities around the world. Comparisons of temperature data from paired urban and rural weather stations suggest that the recent warming trends are due to the urban heat island effect rather than to changes in regional weather. For example, data from thirty-one California cities show a warming rate of 0.4°C (0.7°F) per decade since 1965 (Akbari et al. 1992). Additionally, scientists project a "greenhouse" warming rate of about 0.3°C (0.5°F) per decade, which could exacerbate urban heat-island effects. This section reviews some of the significant economic and social implications of continued urban warming.

Peak cooling loads occur on the warmest days when air conditioning demand is greatest. They increase by about 1 percent for every increase in temperature of 10°C in U.S. cities with populations larger than one hundred thousand. Approximately 3 to 8 percent of the current U.S. electricity used for air conditioning is needed to compensate for the heat island effect caused by an increase in city temperatures of about 1° to 2°C since 1950. Scientists at the Lawrence Berkeley Laboratory (Akbari et al. 1988) estimate total national costs for offsetting the effects of summer heat islands on electricity at about $1 million per hour or more than $1 billion per year (5 percent of total air-conditioning costs).

Urban heat islands can contribute to global warming because warmer temperatures result in greater demands for cooling. Coal-burning power plants release about 0.45 kilograms (1 pound) of carbon per kilowatt-hour of electricity they generate. Therefore, mitigating urban heat islands can indirectly reduce carbon dioxide emissions at power plants and concentrations of atmospheric CO_2.

Large-scale plantings of urban trees and the use of light-colored surfaces can conserve about 2 percent of the total U.S. carbon production (Akbari et al. 1988).

Concentrations of urban ozone are increased by increases in ambient temperature (Cardelino and Chameides 1990). One study found that the incidence of smoggy days increased by 1 percent for each degree Celsius increase in temperature (Akbari et al. 1992). Because many large cities have smog problems and smog concentrations are sensitive to small increases in temperature, controlling urban heat islands is one means of improving air quality.

Urban heat islands can have numerous other adverse effects on the physical and psychological well-being of city dwellers. Heat-aggravated illness and death are related to increased cardiovascular diseases that weaken resistance to heat. Unnaturally high heat loads can directly and indirectly reduce life expectancy (Weihe 1986).

Mitigation of Urban Heat Islands

The thermal behavior of cities is largely a by-product of urban morphology or, more specifically, the composition and three-dimensional structure of materials that constitute the urban canopy layer. This section reviews mitigation potential and the problems associated with urban geometry, surface color, and vegetation, elements that are commonly manipulated during the development process.

Urban geometry. Oke (1988b) stated that designs for street canyons in high- and mid-latitude cities should (1) maximize shelter from wind for pedestrian comfort, (2) maximize dispersion of air pollutants, (3) maximize urban warmth to reduce the need for space heating, and (4) maximize solar access. For warm-climate cities, objective 3 could be modified to read: minimize urban warming to reduce the need for space cooling. In either case, there can be a conflict between objectives promoting shelter, which imply narrow streets and compact forms, and objectives promoting dispersion, which imply separation and low building density. No design can maximize all objectives but it is possible to meet all objectives to a degree that is minimally acceptable.

Oke's analysis of urban geometry in relation to air-flow patterns, radiation exchange, and thermal behavior resulted in a range of canyon geometries and building densities that achieve a "zone of compatibility" wherein the worst aspects of not providing shelter, dispersion, warmth, or solar access are avoided. For instance, aspect ratios (building height/canyon width) of 0.4 to 0.6 and building densities (roof area/total surface area) of 0.2 to 0.4 satisfied all four objectives. The deep canyons and high building densities of many European cities result in geometries that more frequently fall within Oke's zones of compatibility compared to those many North American cities whose settlement patterns are more dispersed (Oke 1988b).

Numerical models and computer simulations have been used to evaluate the climatic impacts of urban design features. Arnfield's (1990) calculations of irradiances on urban canyon walls and floors provide quantitative guidelines on street geometry. He found that irradiance values generally were smaller for canyon walls than for floors, and that controlling canyon-floor irradiance was more critical at lower latitudes because of higher solar angles. North-south street orientation provided less summer and more winter canyon-floor irradiance than east-west street orientation, though the latter provided more favorable seasonal irradiance on canyon walls. Using their Cluster Thermal Time Constant model, Swaid and Hoffman (1990) found that in hot regions, cities with north-south streets have weaker urban heat-island intensities (0.8°C) than cities with east-west streets. Their sensitivity analysis illustrated how increasing aspect ratios (canyon height to width) results in a stronger daytime "cool island" due to increased shade, and a weaker nocturnal heat island due to appreciably reduced maximum temperature.

Although manipulation of aspect ratio, building density, and street orientation might reduce urban heat island effects substantially, the opportunity to exercise this control is limited to new town planning and redevelopment projects. The urban geometry of most cities is well established and large-scale change is unlikely. In these circumstances, the use of vegetation and light-colored surfaces are more feasible mitigation options.

Light-colored surfaces. Increasing the albedo of building surfaces by whitewashing reduces solar heat gain and resulting heat storage. Additionally, increasing the albedo of large areas of the city by using light-colored dyes or sand in paving materials will lower air temperature by reducing the absorption of shortwave radiation. According to data from computer simulations for Sacramento, California, a temperature drop of one to four degrees Celsius can be achieved by increasing citywide albedo from 25 to 40 percent (Taha et al. 1988). Whitewashing residential buildings (albedo change from 9 to 70 percent) reduced annual cooling energy 19 percent and peak cooling demand by 14 percent (Taha et al. 1988). The combined effects of modifications in urban and building albedo resulted in savings of up to 62 percent for annual cooling and 35 percent for peak cooling. However, because the effect of the increased solar load on buildings from lighter surroundings was not incorporated in the model, predicted savings may exceed actual savings.

Albedo modification is a promising strategy for heat island mitigation because of the relatively quick payback period. Once light surfaces are applied, benefits begin to accrue. Many surfaces need to be recoated, so there is the opportunity for large-scale implementation. Using data for Sacramento, Akbari and others (1987) calculated that urban albedo could be increased from 20 to 30 percent by changing the albedo of rooftops (15 to 40 percent) and streets (14 to 30 percent). Increased glare and the desire of some designers and property owners for unrestricted color selection are potential obstacles to implementation. Also, some materials (e.g., tar roofing) do not lend themselves to painting and others may require regular recoating and cleaning (Akbari et al. 1987). Large-scale tree planting could counteract the effectiveness of color change by shading light surfaces and reducing citywide albedo. The extent to which the energy absorbed in this way would influence urban warming depends on its partitioning into latent and sensible heat fluxes.

Vegetated surfaces. The energy-saving potential of trees and other landscape vegetation has been documented (Heisler 1986; Meier 1991; Parker 1990). Vegetation can mitigate urban heat islands directly by shading heat-absorbing surfaces, and indirectly through evapotranspirational (ET) cooling. In a review of studies that measured temperature reductions, Meier (1991) reported that vegetation consistently lowered wall surface temperatures by about seventeen degrees Celsius and reduced air-conditioning costs by 25 to 80 percent. The extent to which measured reductions in surface temperature and savings in cooling costs can be attributed to direct building shade versus ET cooling is not clear. In most circumstances, the impact of one or several trees on ambient temperatures and cooling load are small compared to the shading effect. Cool air produced in the tree crown is dissipated by the much larger volume of air moving through the tree. However, large numbers of trees and expansive greenspaces can reduce local air temperatures by one to five degrees Celsius, and the advection of this cool air can lower the demand for air conditioning (O'Rourke and Terjung 1981; Oke 1989).

Computer simulations (Huang et al. 1987) of the cooling effects of three trees around a residential home in Sacramento showed that shade alone reduced annual and peak cooling energy use by 16 and 11 percent, respectively. The combined effects of shade and ET cooling resulted in savings of 53 and 34 percent, respectively. Thus, ET cooling accounted for 70 percent of the annual energy savings and 68 percent of the peak savings. This finding may reflect the modeling assumption that ET is not limited by soil moisture. Trees in urban environments can experience drought stress due to limited soil moisture and large heat gains from absorbed and reflected radiation (Oke 1989). Theoretically, for stressed plants ET cooling is least during mid- to late afternoon when water vapor deficits are greatest and stomata are closed. Therefore, actual ET cooling effects may be less than projected, especially with regard to peak savings. The relative importance of ET cooling has been disputed by Lowry (1988), who calculated an ET cooling rate of 0.3°C per hour and a sensible heating rate of 1.0°C per hour along ten meters of street canyon containing six mature trees.

The selection and location of trees around buildings to minimize the peak demand for air conditioning is particularly important to electric utilities because the marginal cost of peak power is generally 20 to 30 percent greater than the cost of base load power (about $0.10 versus $0.075 per kilowatt-hour). Parker (1987) described a methodology for determining the optimal location of vegetation to reduce peak loads. "Peak load landscaping" requires that plants be positioned to provide maximum shade several hours before the time when the local utility typically experiences peak demand. In south Florida peak demand usually occurs between 5 and 6 P.M. in early August when temperatures are warmest. To account for the delay before heat on the exterior walls transfers inside and affects the air-conditioning load, Parker recommended plantings for maximum shading of south- and west-facing surfaces between 3 and 4 P.M. in early August. Large-scale implementation of "peak load" landscaping also could reduce urban heat-island intensity because this intensity often is greatest (6 to 11 P.M.) during the period of peak demand for most summer-peaking electric utilities (3 to 10 P.M.). How effectively this landscaping can "shave" and shift peak demand for electricity depends on factors such as the utility's demand profile and regional climate.

Table 2 Summary of Impacts of Vegetation on Electricity Use in Arizona Annual Cooling
Savings (kWh) and Peak Demand Savings (kW)

Study	House Type	Planting Type	Shade		ET Cool		Total/Tree	
			kWh	kW	kWh	kW	kWh	kW
Huang et al. 1987	143 m² wood Phoenix	1 tree west side 50 m²	208	0.47	665	0.33	873	0.80
McPherson & Dougherty 1989	137 m² wood Tucson	1 olive tree west side 46 m²	292	0.39	–	–	292	0.39
McPherson 1990b	137 m² wood Phoenix and Tucson	xeriscape 8 trees 50 shrubs	1673 1353	0.97 0.94	(Phoenix) (Tucson)		209 169	0.12 0.12
McPherson 1991	Tucson	500,000 mesquite	61	–	277	–	338	–

Additional measurements and modeling studies are needed to quantify the potential savings in annual and peak load electricity to be gained through landscaping.

Water is becoming increasingly expensive in many cities, and the cooling value of
trees that use little water has been questioned in desert cities. Shading with little ET
cooling, as with desert trees, results in the conversion of most absorbed radiant energy into sensible heat that warms the air. However, because net storage heat flux
can be substantial in hot, arid cities (Grimmond & Oke 1990), shading alone can reduce urban heat-island intensity by reducing surface temperatures and the amount
of energy stored and later reradiated from paving and buildings (Swaid and Hoffman 1990). Findings from a study using quarter-scale buildings (McPherson et al.
1989) suggest that ET cooling from turf can provide cooling savings equivalent to
the savings gained from the dense shade of shrubs and trees requiring little water.
These results indicate that the microscale effects of vegetation in hot, arid regions
can be more significant than was commonly thought. Landscaped mini-oases with
small turf areas that provide ET cooling and desert trees that shade buildings can
effectively balance the need to conserve energy and water (McPherson 1990b).

Results of several field studies and computer simulations designed to quantify
the effects of trees on electric use in Arizona are summarized in table 2. Annual
cooling energy and peak demand savings ranged from 169 to 873 kilowatt-hours
and 0.12 to 0.80 kilowatts, respectively. Other findings suggest that:

(1) Shading of windows and west-facing walls provides the most savings in cooling energy (McPherson and Dougherty 1989).
(2) On trees selected for shade, crown shape can be more important than crown density (McPherson and Dougherty 1989).
(3) Annual water costs for certain tree species can be twice as much as the energy savings from their shade (McPherson and Dougherty 1989).
(4) Energy and water prices determine the extent to which it is economical to substitute ET cooling for electric air conditioning (McPherson and Woodard 1990).
(5) Effects of tree shade on winter heating demand can be substantial (Thayer and Maeda 1985; Heisler 1986; McPherson et al. 1988).

The Arizona Corporation Commission (1990) recommended that utilities fund the development of consumer guides on energy-efficient landscaping and programs offering rebates for tree planting. These recommendations were based upon the results of a benefit-cost analysis that found that the present value of net benefits for planting 180,000 trees is $2.9 million. The analysis assumed planting costs of $45 per tree, annual water costs of $4 to $6 per tree, a 7 percent discount rate, and a twenty-year planning horizon. Each tree was assumed to shade the west-facing wall and provide annual and peak savings of 250 kilowatt-hours and 0.33 kilowatts, respectively, after the fifth year. The study showed that trees were an economical conservation measure because they met the need for electric energy services at a cost lower than the cost of generating electricity.

A benefit-cost analysis (fig. 1) for the planting of 500,000 trees in Tucson, Arizona, projected net benefits of $236.5 million for the forty-year planning horizon (McPherson 1991). The benefit-cost ratio and the internal rate of return for all trees were 2.6 and 7.11, respectively. Yard trees provided the highest rate of return (14 percent) and street trees the lowest (2 percent). Average annual costs and benefits per tree were estimated as $9.61 and $25.09, respectively. Cooling savings provided the greatest benefits, and removal costs were the largest management expense. Projected average annual benefits in cooling energy per tree were 61 kilowatt-hours ($4.39) for shade and 227 kilowatt-hours ($16.34) for ET cooling. Carbon savings averaged 185 kilograms (408 pounds) annually per tree. An important conclusion was that investment in tree planting by the public sector may be warranted because economic, environmental, and aesthetic benefits extend beyond the site where individual trees are planted.

The value of trees for energy conservation has also been evaluated at the national level. Projections from computer simulations indicate that 100 million mature trees in U.S. cities (three trees for every other single-family home) could reduce energy use for heating and cooling by 30 billion kilowatt-hours and reduce CO_2 emissions by as much as 8 billion kilograms (9 million tons) per year (Akbari et al. 1988). Increasing vegetation was more cost-effective in mitigating heat island effects than other fuel-saving measures such as using energy-efficient appliances and cars.

Although the potential for planting trees in U.S. cities is great and considerable mitigation of urban heat-island effects is possible, there are problems associated with tree planting. First, trees can be a serious liability if they become a public hazard, interfere with aboveground or belowground utilities, or require excessive maintenance. Second, trees can adversely affect the urban climate by blocking solar access in winter, by trapping pollutants within the urban canopy layer and by increasing aerodynamic roughness, thereby reducing country-city air flow and convective heat loss. Third, trees can be relatively expensive to plant and slow to provide a return on investment. Fourth, increased tree planting can increase the amount of pollen that affects allergy sufferers, the use of scarce water supplies, and the amount of solid waste that goes into landfills. However, these problems can be minimized through careful planning, wise selection of species, and designs that fully use the hydrologic, ecologic, atmospheric, restorative, and aesthetic benefits that vegetation can provide.

Figure 1   Projected average annual costs and benefits from proposed planting of 500,000 desert trees during 1990 to 1995 in parks, yards, and streets of Tucson, Arizona (from McPherson 1991)

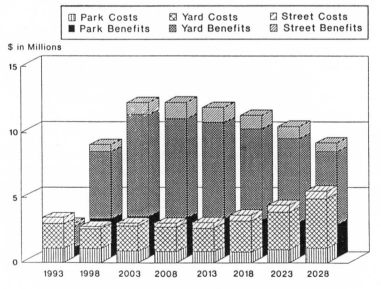

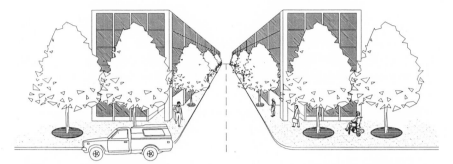

Figure 2   Tall, narrow deciduous street trees shade the canyon walls and sidewalks, reduce downdrafts, and provide ET cooling. Pollution dispersion by down-canyon breezes and vertical mixing of buoyant air is achieved because the crowns do not cover the entire canyon. Bare branches in winter enhance solar heat gain.

Design Examples

Several design examples illustrate how climatological information is applied to urban heat-island mitigation in a northern hemisphere city (approximately thirty degrees north latitude) with a hot climate. These examples focus on the use of vegetation in situations that are typical of urban canyons, parking lots, parks, and single-family residential neighborhoods. Each design addresses objectives defined for street canyons and restated here for hot-climate cities.

1) Promote summertime cooling and the conservation of air-conditioning energy by reducing irradiance and increasing heat loss through latent heat flux, convection, and longwave radiation.

2) Allow for the dispersion of air pollutants by promoting down- and cross-canyon circulation and mixing with air moving above the city.

3) Provide for solar access during the winter by increasing irradiance.

4) Shelter pedestrians from extreme winds, turbulence, and downdrafts near buildings.

Street Canyons

Various techniques have provided a salubrious street-canyon environment by balancing the needs for solar protection and atmospheric dispersion (fig. 2). A high aspect ratio reflects the importance of reducing the irradiance of canyon floors at lower latitudes. Light-colored roof, wall, and paving surfaces further reduce the amount of absorbed radiation. Tall, narrow trees reduce canyon wall irradiance, shade sidewalks, and lessen downdrafts at the base of buildings. Down-canyon ventilation is not entirely obstructed since the trees do not cover the street. This arrangement also allows for some longwave heat loss from street to sky and perhaps for pollution dispersion through the ascension of hot air parcels. Amply irrigated deciduous trees lower summertime temperatures through ET cooling and provide increased winter irradiance when they drop their leaves. The trees also furnish other benefits such as absorbing gaseous pollutants and carbon dioxide, emitting oxygen, intercepting particulates, reducing noise, and enhancing scenic beauty. On the negative side, the trees reduce convective heat loss from the building skin and pollution dispersion from cross-canyon circulation.

Parking Lots

Parking lots are ubiquitous features of commercial strips and shopping malls that surround the urban core. Issues of visibility, safety, screening, and access are central to traditional parking-lot design. The design in figure 3 incorporates these concerns as well as the needs for shade, buffering, and water harvesting (Beatty 1990). Trees are aligned in north-south rows to shade as much pavement as possible during the summer. Dense, broad-spreading tree crowns increase the amount of shade. Ample growing space and soil moisture enhance tree survival, growth, and ET cooling effects. Trees are pruned high enough for safe visibility and truck clearance. Pruning also promotes subcanopy circulation for cooling and pollution dispersion. Rainfall runoff drains into the buffer plantings along the perimeter, where it is detained. These buffer plantings exhibit structural diversity, with a variety of species in several strata. Harvested rainfall supplements irrigation, and the plantings function to reduce sound, particulates, and gaseous pollutants. Vertical accents articulate entryways to enhance the safety and legibility of the parking lot.

Urban Parks

Parks can be important sources of fresh cool air in the city. The park in figure 4 is designed to increase nocturnal cooling and the advection of cool air into the warmer surrounding neighborhood. Much of the park is well-irrigated turf without trees. Transpiring turf shades the ground and cools the air. Although trees in turf can increase the ET cooling effect, they also can reduce radiant heat loss to the sky and convection/advection. This trade-off has not been studied sufficiently to permit us to know whether the addition of trees provides net benefits. Multilayered plantings of drought-tolerant species create a buffer along berms that define the park perimeter. The use of native-looking plants in forms that reflect the

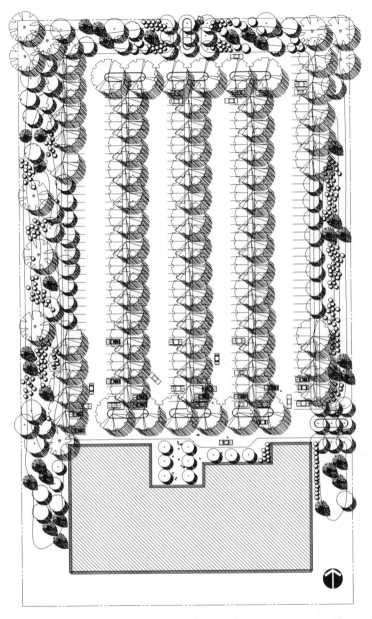

Figure 3   Trees in the parking lot are arranged in north-south rows to provide maximum shade on pavement during the summer. Sufficient soil volumes and irrigation promote rapid growth and ET cooling. Stormwater runoff is harvested in basins along the periphery, where structurally diverse plantings filter particulates, intercept rainfall, and screen the parking lot.

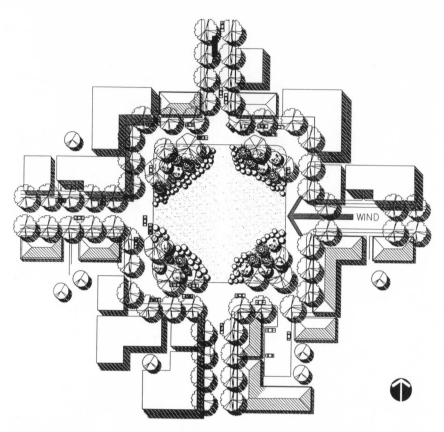

Figure 4   The open turf area in this urban park is a source of cool air that is carried into the surrounding area by prevailing breezes. Buffer plantings along the park boundary reflect the complex structure of the region's native plant associations.

structure of native plant communities promotes a sense of place. The buffer planting is multifunctional and an important symbol of "naturalness" in the heart of the city. The berms and buffer plantings disappear at street intersections to facilitate the flow of cool air through downwind openings and into adjacent street canyons.

Residential Streets

Most residential streets in the United States have paving that is too wide and planting strips that are too narrow. The design in figure 5 "reclaims" the right-of-way for people, plants, and animals. Narrowed traffic lanes force cars to travel slowly through residential neighborhoods yet allow access for emergency vehicles. Parking for residents and guests is provided along intrablock alleys. Large roadside trees shade the pavement and promote pedestrian travel within a more comfortable microclimate. Reclaimed irrigation water and stormwater harvested off the surrounding areas nourish a complex association of plants and animals. Smaller drought-tolerant trees shade adjacent residences, conserving energy and water.

Conclusion

Although urban climatologists have made many measurements of urban heat-island effects, there are relatively few predictive models for use by those who design

Figure 5 Reclaimed greenspaces along residential streets provide areas for naturalistic plantings that enhance biodiversity by creating riparianlike habitats. Runoff harvested from streets and sidewalks reduces flooding downstream and conserves irrigation water. Small trees that use little water shade nearby buildings.

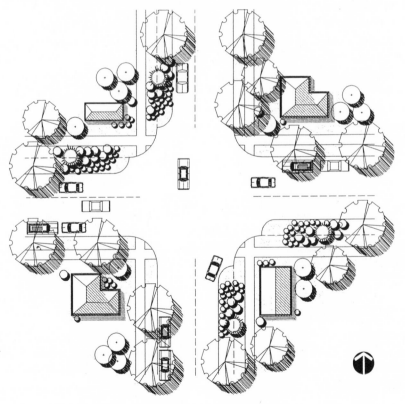

our cities (Oke 1988a). Most existing models do not adequately consider all fluxes of energy and matter or incorporate the spatial and temporal variability within the urban canopy layer (Martien et al. 1990). There is need for numerical models that are based on the physical processes at work in cities (Grimmond and Oke 1986). The output from microclimatic or canopy layer models that function at the scale of a homesite could be input for mesoclimatic models that integrate microscale effects over the area of a neighborhood or city. Once these models have been validated in different climatic regions, they should be simplified and integrated with existing geographic information systems to facilitate implementation by urban planners.

Studies are needed to determine the most cost-effective method for obtaining land-surface information for modeling purposes. Remotely sensed information seems appropriate for mesoscale modeling but may not provide the resolution needed for microscale simulation. Field surveys might be needed to identify plant species and size as well as building construction types, and to collect other detailed information. This will require the development of appropriate sampling and survey techniques.

Environmental planners are asking many practical questions related to urban heat-island mitigation that we cannot answer now. For instance, policy-makers in Arizona want to know the trade-offs between ET cooling and water demand. Elsewhere, utility and urban planners want to know how much savings in cooling energy can be achieved by increasing the amount of urban vegetation by a specified amount. Designers are asking how best to locate and manage vegetation in parks, streets, and residential areas to improve urban climate. Urban foresters want to know which tree species will provide the greatest long-term net benefits.

Clearly, there is a need for further development of urban climate models, as well as for climatological and urban forestry research that can be used to validate and verify the models. Equally important is the development and application of "user-friendly" planning and design tools. The cooling of our urban heat islands depends on the timely development and implementation of reliable predictive models, and the guidelines, regulations, and incentives that can be generated from the information they provide.

Acknowledgments

I appreciate reviews of earlier versions of this manuscript by Sue Grimmond (Indiana University), Hashem Akbari (Lawrence Berkeley Laboratory), Rich Grant (Purdue University), and Craig Johnson (Utah State University). Steve Wensman provided assistance with the illustrations.

References

Akbari, H., H. Taha, P. Martien, and J. Huang. 1987. Strategies for reducing urban heat islands: Savings, conflicts, and city's role. In *Proceedings of the First National Conference on Energy Efficient Cooling.* San Jose, CA.

Akbari, H., J. Huang, P. Martien, L. Ranier, A. Rosenfeld, and H. Taha. 1988. The impact of summer heat islands on cooling energy consumption and CO_2 emissions. In *Proceedings of the 1988* ACEEE *summer study on energy efficiency in buildings.* Washington: American Council for an Energy Efficient Economy.

Akbari, H., A. Rosenfeld, and H. Taha. 1990. Recent developments in heat island studies: Technical and policy. In *Controlling Summer Heat Islands,* ed. K. Garbesi et al. Berkeley: Lawrence Berkeley Laboratory.

Akbari, H., S. Davis, S. Dorsano, J. Huang, and S. Winnett. 1992. *Cooling our communities: A guidebook on tree planting and light-colored surfacing.* Washington: U.S. Environmental Protection Agency.

Arizona Corporation Commission. 1990. *Resource planning staff report.* Phoenix: Utilities Division, Arizona Corporation Commission.

Arnfield, A. J. 1990. Street design and urban canyon solar access. *Energy and Buildings* 14(2): 117–31.

Beatty, R. A. 1990. Planting guidelines for heat island mitigation and energy conservation. In *Controlling summer heat islands,* ed. K. Garbesi, H. Akbari, and P. Martien. Berkeley: Lawrence Berkeley Laboratory.

Cardelino, C. A., and W. L. Chameides. 1990. Natural hydrocarbons, urbanization, and urban ozone. *Journal of Geophysical Research* 95(D9): 13971–79.

Chandler, T. J. 1965. *The climate of London.* London: Hutchinson.

Duckworth, F. S., and J. S. Sandberg. 1954. The effect of cities upon horizontal and vertical temperature gradients. *Bulletin of the American Meteorological Society* 35:198–207.

Goward, S. N. 1981. Thermal behavior of urban landscapes and the urban heat island. *Physical Geography* 2(1): 19–33.

Grimmond, C. S., and T. R. Oke. 1986. Urban water balance 2: Results from a suburb in Vancouver, B.C. *Water Resources Research* 22:1404–12.

———. Preliminary comparisons of measured summer suburban and rural energy balances for a hot dry city, Tucson, Arizona. *Proceedings of the International Symposium on Urban Climatology, Air Pollution and Planning in Tropical Cities,* 25–30 November, 1990. Guadalajara, Mexico.

Heisler, G. 1986. Energy savings with trees. *Journal of Arboriculture* 12(5): 125.

Herrington, L. P., G. E. Bertolin, and R. E. Leonard. 1972. Microclimates of a suburban park. In *Proceedings of a Conference on Urban Environment and the Second Conference on Biometeorology,* Boston: American Meteorological Society.

Huang, J., H. Akbari, H. Taha, and A. Rosenfeld. 1987. The potential of vegetation in reducing summer cooling loads in residential buildings. *Journal of Climate and Applied Meteorology* 26:1103–6.

Landsberg, H. E. 1981. *The urban climate.* New York: Academic Press.

Lowry, W. P. 1988. *Atmospheric ecology for designers and planners.* McMinnville, OR: Peavine Publications.

Martien, P., H. Akbari, A. Rosenfeld, and J. Duchesne. 1990. Approaches to using models of urban climate in building energy simulation. In *Controlling Summer Heat Islands,* ed. Garbesi et al. Berkeley: Lawrence Berkeley Laboratory.

McPherson, E. G., L. P. Herrington, and G. Heisler. 1988. Impacts of vegetation on residential heating and cooling. *Energy and Buildings* 12:41–51.

McPherson, E. G. and E. Dougherty. 1989. Selecting trees for shade in the Southwest. *Journal of Arboriculture* 15(2): 35–43.

McPherson, E. G., J. R. Simpson, and M. Livingston. 1989. Effects of three landscape treatments on residential energy and water use in Tucson, Arizona. *Energy and Buildings* 13:127–38.

McPherson, E. G. 1990a. Creating an ecological landscape. In *Proceedings of the Fourth Urban Forestry Conference,* ed. P. Rodbell. Washington: American Forestry Association.

———. 1990b. Modeling residential landscape water and energy use to evaluate water conservation policies. *Landscape Journal* 9(2): 122–34.

———. 1991. Economic modeling for large-scale urban tree plantings. In *Energy Efficiency and the Environment: Forging the link.* Washington: American Council for an Energy Efficient Economy.

McPherson, E. G. and G. C. Woodard. 1990. Cooling the urban heat island with water-and-energy efficient landscapes. *Arizona Review,* (Spring):1–8.

Meier, A. K. 1991. Measured cooling savings from vegetative landscaping. In *Energy Efficiency and the Environment: Forging the link.* Washington: American Council for an Energy Efficient Economy.

Oke, T. R. 1982. The energetic basis of the urban heat island. *Quarterly Journal of the Royal Meteorological Society* 188(455): 1–24.

Oke, T. R., ed. 1986. *Urban climatology and its application with special regard to tropical areas.* Geneva: World Meteorological Organization.

Oke, T. R. 1987a. *Boundary layer climates.* New York: Methuen.

Oke, T. R., ed. 1987b. *Modeling the urban boundary layer.* Boston: American Meteorological Society.

Oke, T. R. 1988a. The urban energy balance. *Progress in Physical Geography* 12:471–508.

———. 1988b. Street design and urban canopy layer climate. *Energy and Buildings* 11: 103–13.

———. 1989. The micrometeorology of the urban forest. *Philosophical Transactions of the Royal Society of London* 324: 335–49.

O'Rourke, P. A. and W. H. Terjung. 1981. Urban parks, energy budgets, and surface temperatures. *Archives for Meteorology, Geophysics and Bioclimatology* 29: 327–44.

Parker, J. H. 1987. The use of shrubs in energy conservation plantings. *Landscape Journal* 6(2): 132–9.

———. 1990. The impact of vegetation on air conditioning consumption. In *Controlling Summer Heat Islands,* ed. Garbesi et al. Berkeley: Lawrence Berkeley Laboratory.

Rowntree, R. A. 1984. Forest canopy cover and land use in four eastern United States cities. *Urban Ecology* 8:55–67.

Swaid, H. and M. E. Hoffman. 1990. Climatic impacts of urban design features for high- and mid-latitude cities. *Energy and Buildings* 15(4): 325–36.

Taha, H., H. Akbari, A. Rosenfeld, and J. Huang. 1988. Residential cooling loads and the urban heat island-the effects of albedo. *Building and Environment* 23(4): 271–83.

Thayer, R. L. and B. Maeda. 1985. Measuring street tree impact on solar performance: A five climate computer modeling study. *Journal of Arboriculture* 11:1–12.

Tuller, S. E. 1973. Microclimatic variations in a downtown urban environment. *Geografiska Annaler* 54(3–4): 123–35.

Tyson, P. D., M. Garstang, and G. D. Emmitt. 1973. *The structure of heat islands.* Occasional paper no. 12, Department of Geography and Environmental Studies, University of Witwatersrand. Johannesburg.

Weihe, W. H. 1986. Life expectancy in tropical climates and urbanization. In *Urban Climatology and its Applications with Special Regard to Tropical Areas,* ed. T. R. Oke. Geneva: World Meteorological Organization.

Wildflower Meadows as Sustainable Landscapes

Jack Ahern and Jestena Boughton

Introduction

In the industrialized world of today, nature often has a limited role in the urban landscape. Contemporary life has largely disassociated humans from their native physical environment. Some view this as a sign of sophistication and urbanity, others look at the fouled environment of our cities and believe it is time for rethinking the role of nature in the city. Frequently, the "nature" that is present in the urban environment exists as a superficial entity dependent on humans for nonsustainable inputs of materials, energy, and labor. The urban landscapes that are often admired such as rose gardens, emerald green golf courses, and theme parks represent nonsustainable phenomena heavily subsidized with labor and agrochemicals. Up to 80 percent of the species used in these landscapes are exotics, from other parts of the country or other continents. The love of the exotic permeates current landscape aesthetics. Our culture has been trained to appreciate the exotic and to shun the common or indigenous. This is a fundamental issue to be confronted in developing a new aesthetic that is consistent with sustainability, one that expresses uniqueness and local character. David Northington, the director of the National Wildflower Research Center at Austin, Texas, has spoken on this issue:

> We have lacked confidence in our taste, and in a sense, in our own flora. We need to break with some of the old, fairly ridiculous fashions, such as trying to turn every backyard or park into an English formal garden. One of the things we're doing . . . is helping people to discover that it makes ecological sense, and is aesthetically challenging to use native species and create settings for them that reflect the natural realities of our land rather than those of England, Italy, or Japan. (Northington in Gilbert 1987, 45)

Sustainability is a concept that has great relevance to landscapes. *Sustainability* is defined as "a condition of stability in physical and social systems achieved by accommodating the needs of the present without compromising the ability of future generations to meet their needs" (World Committee on Environment and Development 1987). Building on this concept, we define *sustainable landscapes* here as those that exist with a minimum of nonsolar inputs of energy, materials, or labor. They are productive landscapes in that they accumulate more organic material than they consume. They provide forage and habitat for wildlife species. They are regionally appropriate, supportive of the native biodiversity of plants and animals in their locale. Finally, sustainable landscapes are responsive to human needs. They foster human understanding of the natural history and natural processes of a place. They also provide delight and address functional needs; if they do not, they enjoy limited support, especially in urban areas.

Sustainable landscapes take time to develop. To think otherwise denies the importance of ecological growth and change, and treats the landscape as an inanimate element for visual pleasure and functional necessity only. This attitude is perhaps the larger reason that many attempts at establishing wildflower meadows are terminated prematurely because of unrealistic expectations, or impatience with an establishment period that is much longer than that for conventional landscapes such as turf grass. The landscape industry thrives on an "instant landscape" mentality that accelerates ecological growth and development with heavy subsidies of labor and chemicals and with monocultures of genetically altered plants. Paradoxically, the landscape industry has become a major obstacle to the acceptance of a new landscape aesthetic founded explicitly on sustainable native species and natural processes.

Wildflower Meadows as Sustainable Landscapes

Wildflower meadows are sustainable landscapes. Defined here as diverse communities of native and naturalized forbs and grasses, meadows enhance awareness of seasonal changes and expose surrounding vistas. Meadows that consist of native vegetation also provide important habitat and food for wildlife. An established meadow is resistant to weed invasions and has a greatly reduced need for maintenance as compared with the needs of turf grass lawns (National Wildflower Research Center 1989; Martin 1986). Proper management can sustain a desirable species mix and an attractive appearance.

People can learn much about the interactions of plants, wildlife, and the land from observing a meadow over time. Meadows change. Some species stay, others leave and are replaced. Meadows are a constantly changing mosaic of species in an ongoing process of adaptation and survival. Meadows can be managed to reveal and teach more about the environment through species selection, interplanting, and careful timing of mowing or burning.

Wildflower meadows that replace conventional turf grass offer three principal benefits: ecological, economic, and aesthetic. The extent to which these benefits are realized is a function of numerous factors, including species selection, planting site location, establishment success, and management. The following sections will elaborate on these potential benefits.

Ecological Benefits

Maintaining a diversity of habitat types has long been recognized as important to environmental health and quality (Gore 1992, 139). Urban landscapes managed as meadows contain substantially more diversity than conventional turf-grass landscapes. In addition to offering a greater diversity of plants, meadows provide habitat for many species of wildlife and insects. The fact that routine meadow maintenance requires only an annual or biannual mowing greatly reduces disturbance and creates more valuable wildlife habitat.

Meadows may serve as a vegetational buffer between the intensively managed landscape and the adjacent forest, roadway, or building. Meadow/forest buffers increase the ecological value of remnant woodlands in urban areas by enlarging the effective edge zone around them and giving more protection to the interior habitat. If implemented on a large scale, urban meadows could contribute to a network of linked corridors supporting a diversity of plant and animal life.

Another form of ecological benefit is the reduced environmental impact associated with maintenance of meadows compared with the maintenance of turf grasses. A recent newspaper article proclaimed "the nation's weed-free carpet [turf grass] is soaked in poison." The average American lawn receives four times as much chemical pesticide as any U.S. farmland (Pollan 1991). Managing turf grass also has a significant impact on water quality, because pesticides, oil and gasoline, lead, and sediments are contributed to runoff that traverses the turf. A water quality index of selected land uses found that golf courses had the highest water pollution (Fabos 1985, 26), due to the intensive management of the large turf areas involved. Since the maintenance of wildflower meadows does not involve the use of pesticides or fertilizers, a direct reduction in this form of nonpoint-source pollution is possible. Reductions of other pollutants can also be realized through the ability of wildflowers to trap and filter air-borne pollutants. Plants trap these pollutants on their leaves, stems, and trunks. The efficiency of this "trapping" increases directly in proportion to the plant's total surface area (Spirn 1984). Thus, a two-to-three-foot-tall growth of wildflowers will be substantially more efficient at trapping pollutants than a three-to-six-inch-tall growth of turf grass. This ability is particularly significant in urban situations and for landscapes adjacent to highways. Further, once the pollutants are trapped by the vegetation, they are more likely to leach into the soil than to run off into surface waters, because the runoff associated with the wildflower meadow is reduced as compared with turf runoff (Leopold and Dunne 1978). As soil is an effective sink for airborne pollutants, a substantial improvement of water quality can be accomplished in wildflower meadows as compared with areas under turf grass management.

In the midwestern states, over 99 percent of the original prairie vegetation has been destroyed through plowing. Most of the remaining prairies are located along railroad lines and highways, where they were protected from plowing and grazing. These prairie remnants contain many rare species and are now invaluable ecological reserves where seed stock and models for ecological restoration may be found for future use in other landscape contexts. The protection of existing reserves and the restoration of others represent highly significant benefits provided by wildflowers.

Economic Benefits

The primary economic benefit associated with substituting meadows for lawns in urban areas is the reduced need for mowing. While turf grasses require six to twenty mowings per year (depending on the weather, soil and moisture conditions, and desired appearance), typical maintenance of wildflower meadows involves only one mowing per year. Thus, for every area managed as wildflowers versus turf, a minimum cost savings of 83 percent (five out of six mowings) can be realized. Every year an average of thirty hours for every man, woman, and child in the United States are spent mowing lawns (Pollan 1991). Additional cost savings result from reduced gasoline consumption and reduced water consumption for irrigation.

The Massachusetts Department of Public Works mowed roughly 3,300 acres of roadside turf in 1987 at a cost of approximately $1 million, or $330 per acre based on six mowings per year. For every acre managed as wildflowers versus turf, mowing can be reduced to one annual cut with an annual cost savings of over $280 per acre (Evans 1987). The actual cost savings may differ considerably from this potential one, however, when wildflowers are planted in small "planting beds," because the time and effort to mow around them eliminates most or all cost savings. This is an important consideration in the planning and design of highway and other wildflower meadows.

Wildflower meadows on highway rights-of-way also produce cost savings by controlling snow drifting and soil erosion, and by reducing the need for plant replacement. In several states, highway wildflower meadows have been protected and managed to support growing tourism industries. Many states have officially designated wildflower routes to attract tourists.

Aesthetic Benefits

Wildflowers are often praised for their aesthetic benefits, including color, interesting textures, and indications of seasonal change. Research on environmental preferences has found that while nature is often considered synonymous with the open landscape, wide open spaces are not universally preferred (Kaplan 1984). In fact, landscape scenes that lack spatial definition tend to be relatively unliked. Only when areas include elements that help to differentiate the openness, such as groupings of trees and shrubs, are they preferred. These spatially defined landscapes, often described as parklike or savanna, have been found to invoke high preference in a number of studies. The findings of this research on landscape preference suggest that in order to make urban landscapes more aesthetically appealing, elements that articulate and differentiate the landscape's visual space should be incorporated. Wildflowers and masses of native woody plants are means of achieving this preferred vegetative diversity while simultaneously realizing the ecological and economic benefits previously discussed. Meadows in urban landscapes provide an additional benefit in that they contribute to defensible space by not screening human sight lines. This was the primary reason for establishing wildflowers in New York's Bronx to replace tall-growing exotic trees and shrubs, known locally as mugger cover. Wildflower meadows are also increasingly used in urban contexts as transitions between formally maintained landscapes and relatively unmanaged areas (Blake 1990).

Wildflower meadows may contribute to improved landscape aesthetics by expressing the underlying physical character of a site. For example, in moist or wet areas of the Northeast, wet meadows can be developed that include prominent species such as Joe-Pye weed, swamp milkweed, cup plant, ironweed, and Turk's-cap lily. Such diverse wet meadows graphically reveal the presence of wetlands in the landscape, and thus add visual interest and promote ecological understanding.

Meadows are not universally accepted as beautiful, however. There is often a perception of neglect associated with meadows in urban contexts, particularly if the interface with managed turf areas is poorly considered (Burley and Burley 1991; Hough 1984). Unmown turf often represents neglect and abandonment of responsibilities. Many meadow advocates would recommend maintaining turf in areas under intensive human use or where the appearance of care and control is important. Others have argued that prominent public lawns as at the White House should be removed and replaced with meadows and wetlands to symbolize conspicuously a new paradigm of environmental responsibility (Pollan 1991).

Figure 1   Wildflower meadows can be creatively integrated into many different landscapes to create visual diversity and interest. At Lincoln Park in Lexington, Massachusetts, wildflower meadows, old fields, and wetland communities have been restored on a former landfill site.

A Brief History of the Wildflower Movement

The use of wildflower meadows in designed landscapes in the United States can be traced back as far as Andrew Jackson Downing and Frederick Law Olmsted, Sr., in the mid- to late nineteenth century. While these landscape designers frequently used native plants, including wildflower species, in their work, it was more in the interest of creating the desired romantic scenery than of re-creating a natural community of plants. The first extensive use of wildflower meadows as natural plant communities occurred in the Midwest by Ossian Simonds and Jens Jensen in the late nineteenth century. These designers were leaders of the "prairie landscape architecture" movement, which embraced the concept of designed landscapes modeled after natural landscapes, particularly the prairies of the Midwest. Jensen designed parks, golf courses, cemeteries, and conservation areas for the Chicago Parks Department. He was committed to a prairie landscape aesthetic based on native plants and used the palette of midwestern prairie species masterfully throughout his work (Grese 1988).

The next extensive use of wildflowers on a large scale in the United States was in

Texas in the 1920s, following a period of major highway construction related to the rising popularity of automobiles. Highway crews observed that native species were the first to revegetate disturbed construction rights-of-way. They realized that this native vegetation would be much easier to establish and maintain than green lawns (McCommon 1983). Thus, the use of native wildflower meadows was adopted as a standard model for vegetation management in the infancy of highway construction in Texas. Texas subsequently developed innovative methods of harvesting the large quantities of seed necessary for new plantings by sickle-cutting existing highway meadows. This method provides a seed stock that is adapted to local climatic and soil conditions and contains many of the indigenous species of the area. The highway managers in Texas have learned to coordinate the mowing schedule with the blooming periods of selected species such as the Texas bluebonnet. After the major spring bloom, the meadows are cut, thereby spreading the seeds for regeneration. The highway meadows are a major feature of a statewide series of wildflower festivals. Texas now has over three-quarters of a million acres of meadows on seventy thousand miles of highways. In the 1970s the Texas highway department updated its program and set three primary goals: to reduce maintenance costs, to develop a sound management policy for vegetation that is aesthetically pleasing; and to make the highway right-of-way blend into the surrounding landscape in an unannounced manner. In Texas it is estimated that $32 million are saved per year by encouraging native species because of the attendant reduction in mowing and spraying requirements (Gilbert 1987).

Operation Wildflower is a unique cooperative effort by the Federal Highway Administration (FHA), the National Council of State Garden Clubs, and state highway departments. Under this program the FHA provides the land for the plantings, state highway departments select sites for planting and supply maintenance, and local garden clubs contribute seeds, plants, or funds to purchase planting materials. The goals of Operation Wildflower are to lower costs, provide color, restore native communities, and to protect wildlife. As of 1977 only thirteen states were participating; in 1986 there were thirty-eight, and more recently the total has reached forty-six. Clearly wildflower meadows have earned a permanent place along the nation's roadsides.

A "Roadside Wildflower Task Force" in Minnesota has developed a statewide policy to preserve existing native wildflower populations, restore native wildflowers where appropriate, and educate the public about native wildflowers (Minnesota Roadside Wildflower Task Force 1988). The group's recommendations include goals for preservation, restoration, and education. The suggested policies are based on ecological, economic, aesthetic, and functional benefits. The task force has integrated a broad cross section of representatives of the public and private sectors to develop a truly visionary wildflower meadow program.

The National Wildflower Research Center at Austin, Texas, was established through a donation by Lady Bird Johnson in 1982. The center's mission is to conduct research to promote the use of wildflowers by nurseries, highway departments, landscape professionals, scientists, and amateur gardeners. The center also acts as a clearinghouse for information on wildflowers for general use, and is the only such clearinghouse in the world. The center is committed to the belief that wildflower meadows have a place within the grid of urban and suburban development, inside highway cloverleafs, surrounding suburban office parking lots, in planned residential communities, and in private gardens. In this respect the group's approach differs fundamentally from that of the native plant society, which emphasizes protection and preservation, particularly of rare and endangered species. The center's establishment marked the reemergence of wildflower meadows into the mainstream of contemporary landscape design.

Wildflower Meadow Issues

There is presently much interest in using wildflower meadows in a range of landscape contexts, from highways to office parks to private residences. Many earlier attempts at wildflower meadows have been unsuccessful because of a lack of understanding of the basic issues involved and unrealistic expectations concerning the results. These difficulties are not surprising in light of the fact that successful establishment of wildflower meadows requires an approach to landscape design and management fundamentally different from the current one. This approach begins with a keen understanding of site conditions, which becomes the basis for a landscape design that is then crafted to match the client's expectations in terms of time of establishment, cost, and results desired. The following is a summary of the principal issues involved.

The "Meadow-in-a-Can" Myth

In the early 1980s seed companies aggressively marketed wildflower seed for landscapes throughout the United States. They promised brilliantly multicolored meadows in perpetual bloom without any maintenance. Virtually no mowing, no watering, no fertilizers were needed, according to their claims. It almost sounded too good to be true—and it was. The results from these plantings ranged from sustained success to marginal failure to wholesale replacement of meadows with turf grass. Research has shown that improper site preparation and establishment techniques were the leading factors contributing to unsatisfactory results (Ahern 1990). An additional problem was a conflict between the meadow gardener's expectations and the reality of a meadow planting. The idyllic image of the "meadow-in-a-can" that could simply be broadcast over the soil and watered unfortunately became a widespread aspiration, but it was unrealistic in several respects. First, meadow establishment requires intensive soil preparation, a great deal of time, or both. There simply is no quick method of establishing a stable community of meadow plants. Secondly, the brilliant floral displays of wildflowers in seed catalogs invariably comprise a high percentage of annual species that will flourish the first year but diminish or disappear rapidly in subsequent seasons. Furthermore, many of these species are not remotely wild or even native to many regions of the United States (Burley and Burley 1991). A study by the New England Wildflower Society found that many of the commercially popular "Northeast regional" seed mixes contained 50 percent or fewer of species native to the Northeast (Meyers 1990).

Realistic expectations for wildflower meadows should involve an establishment period of two to three years during which greater-than-usual maintenance is required. When successfully established, the meadow should contain approximately equal percentages of wildflowers and native clump-forming grasses. The aesthetic effect is not analogous to that of a bed of annuals or a perennial border. A meadow is more restrained in its floral display, yet offers far greater aesthetic diversity over the course of the year. The grasses in the meadow are particularly striking visually in fall and wintertime.

The Weed Problem

Proper site preparation and establishment techniques are essential to achieving a stable plant community and an aesthetically successful wildflower meadow. Proper site preparation will permit the establishment of wildflowers while reducing the number of undesirable opportunistic grasses and weeds that can successfully invade the meadows and displace desirable wildflowers and clump-forming grasses.

One of the most common causes of failure in wildflower meadow plantings is weed invasion. This issue is loaded with confusion, beginning with the definition of a *weed*. One person's wildflower is often another person's weed. A useful definition of a weed is a plant that (1) colonizes disturbed habitats, (2) is not a regular member of the original natural community of the geographic area in which it is found, (3) is abundant, at least locally, (4) is noxious, destructive, or troublesome, and (5) has little economic value (Zimmerman in Radosevich and Holt 1984, 10). Weeds have many strategies for competing successfully in herbaceous communities, including a rapid growth rate, prolific seed production, and tolerance of stress. Interestingly, many of the more pernicious weeds that grow in turf grass lawns are not successful in meadows. Plants such as dandelions and crabgrass actually thrive on the mown and fertilized environment of a conventional lawn. In a meadow, these species usually succumb to taller-growing forbs and grasses.

Figure 2   Wildflower meadows are increasingly used in the large corporate landscapes found in suburban office parks.

Unfortunately, there are many species of grasses and forbs that behave as weeds and are undesirable in meadows. In particular, species that reproduce vegetatively may become invasive and displace the desired forbs and grasses. Species such as Canada goldenrod and some perennial rye and fescue grasses are invasive in meadows in the northeastern United States. These species may be holdovers from plantings that preexisted the meadow, or species that were patiently waiting in the soil as dormant seeds for the right conditions. Cultivation to prepare sites for meadow planting often exposes a large volume of dormant weed seeds from the seed bank in the soil. These seeds then enjoy the care lavished on the wildflower species that are planted. By the time most wildflower gardeners realize they have a problem, it is too late to remove the weeds easily.

The solution to the weed problem is not complex but does require labor, patience, and/or use of herbicides. The key is to control the weeds before planting the wildflower meadow. This can be achieved through repeated cultivation, or multiple applications of herbicides to deplete the seed bank of weed seeds (Ahern 1990). The greatest defense against weed invasion is to establish a vigorous and diverse community of wildflowers and grasses. The removal of clippings from mowing is often recommended to maintain lower fertility levels in the soil and thereby favor wildflowers and grasses over weeds. Particular attention must be paid to areas of disturbance, such as eroded slopes or soils compacted by foot traffic, which may provide opportunities for weed invasion. Should weeds become a problem later in the establishment period, they may be removed in small meadows by manual pulling, or selectively sprayed with herbicide by someone capable of identifying the meadow species to be protected.

The Species Selection Dilemma

The composition of species in the wildflower meadow is another issue to consider. Should a meadow contain only species indigenous to the United States or to a specific region, say, the Southeast? Should the meadow include some Eurasian species that have become naturalized in the United States to the point where most people consider them native wildflowers (as with the common ox-eye daisy)?

Perhaps it is useful to consider the two extreme responses to this dilemma. The "purist" approach uses only locally indigenous species, from within a radius of maybe two hundred miles of the planting location. Implementation of this approach requires locally grown seed, as imported seed, even of the proper species, is not adapted to the photoperiod, soil, and climatic factors of the region. This is an issue of serious concern with respect to biodiversity. Some would argue that it is preferable to plant nonaggressive exotic species rather than to import nonregional genotypes that will alter the gene pool of locally native species. Advocates of the purist approach are particularly active in the Midwest, where the ecological significance and uniqueness of the native prairies are highly valued. Seeds are typically collected by volunteers from remnant or restored prairies. This approach tends to optimize the ecological benefits of wildflower meadows, but is difficult to implement on a commercial scale because of the limited supply of locally grown or collected seed.

The other extreme is the "horticultural" approach, which views wildflower meadows primarily as visual elements and selects species to provide an attractive sequence of bloom throughout the growing season. The palette of species is usually limited, in the interest of achieving a significant splash of color at several times during the growing season. Annual species are often included in the mix for additional color. This approach maximizes the aesthetic effect (for certain tastes) at the expense of ecological benefits. Economic benefits are also somewhat reduced, as these plantings tend to be stable for a limited time before they are taken over by weeds. They are frequently replanted, or overplanted every four to six years. Many highway meadows with high visibility are planted in this manner to create a prominent visual effect, which requires large quantities of seed to achieve on a statewide basis.

Of course, in actual application endless combinations and compromises are possible between the purist and the horticultural approaches. The Brandywine Museum in Chadds Ford, Pennsylvania, for example, features on its grounds a wildflower meadow garden that emphasizes native species but includes some common naturalized species. The garden has become a regional center for native flora display and education, and is maintained by a network of volunteers who collect and propagate seeds for the garden. Small amounts of locally collected seed are available for noncommercial meadow gardeners in the Philadelphia region.

The meadow garden at the Garden in the Woods in Framingham, Massachusetts, was conceived to demonstrate the greatest possible diversity of plants native to the eastern United States that have desirable ornamental characteristics. The meadow was established on a small area by planting over three thousand live plants. Native grasses were established by seeding. The meadow provides a continuous bloom from late May through October. It includes many species native to New England but also includes numerous species from the Midwest. The garden requires fairly heavy maintenance to maintain simultaneously a high level of diversity and a striking floral display.

Wildflower Meadow Management

Once a wildflower meadow has become successfully established, management is a relatively straightforward matter. Most of the species in the meadows are annuals or perennials with life expectancies of one to several years. They all have the capability to reproduce through vegetative growth or seed germination. While the individual species may come and go, the overall meadow community demonstrates a dynamic form of stability. The meadow can thus be seen as a constantly changing mosaic of herbaceous forbs and grasses.

If no maintenance were performed on an urban meadow in a temperate climate, it would be likely to become an old field or young forest in a matter of years or decades. To prevent this natural successional tendency, management intervention is needed to stabilize the meadow community and arrest the process of succession at the herbaceous meadow stage. There are three principal ways of achieving this stabilization: mowing, grazing, or burning. While at present only the mowing option is generally considered appropriate in urban situations, the control of meadow succession through grazing and burning is worth consideration (Andersen 1990).

Mowing selectively favors herbaceous plants over woody species, as the latter invest more in producing aboveground biomass that is easily removed by mowing. The mowing can be timed to control which species will persist in a meadow

(Emerson 1990). Mowing plants just before the flowers mature will often exhaust their stored energy and prevent them from setting seed. This technique can be used to control invasive meadow species such as Canada goldenrod. The management objective in most meadow maintenance is usually achieved by an annual mowing at four to six inches. The optimal time is late fall, after the seeds have matured. A delay of the mowing until early spring produces the same control but leaves a standing dormant cover for visual interest and wildlife cover throughout the winter.

Creatively designed edges between mown turf and unmown meadows can enhance recreational use and pedestrian circulation. Mown edges can also maintain a sense of intentional management and control, and address functional needs including litter control and protection from fire. Urban meadows have their place, but they are not a panacea for the urban landscape. Areas under intensive human use, areas adjacent to buildings where fire control is an important consideration, intensively used recreational areas, and areas where the appearance of care and control are important are probably best left as turf. The Dutch have successfully integrated naturalized meadows and carefully mown areas in urban parks to address the diverse needs for urban recreation (Bos and Mol 1979).

The management of the edges of wildflower meadows presents challenges of a different kind. The matrix of turf grass that often surrounds meadows is usually mown six times a year. Over the course of the growing season, the mowing produces an unnaturally abrupt edge that often dominates the impression of the meadow as seen from a highway. In Massachusetts, an alternative scheme of "stepped mowing" has been designed to resolve this problem. Under this method, each time the turf is cut, the mow line is sequentially moved away from the meadow, producing a "stepped," or graded, edge by the end of the season. This method has been found to be an effective way to enable existing wildflower plantings to increase incrementally in size without any additional cost or effort.

Grazing is employed in the management of urban grasslands in Europe and Canada. A petroleum company in Toronto has used sheep to graze grasslands around its petroleum tanks for over ten years. This company has found that one sheep can graze three to four acres in closely cropped and fertilized condition, and eliminate the need for mowing equipment (Hough 1984, 156). Burning is a technique also used in maintaining wildflower meadows. Controlled burns simulate naturally occurring fires, which are an integral part of the ecology of grassland ecosystems. Grasses, with a high percentage of biomass underground, are less affected by fire than woody species. After a fire, increased light and nutrient availability actually cause grasses and wildflower forbs to flourish. Periodic burning is an essential maintenance activity for true prairies, because it selectively

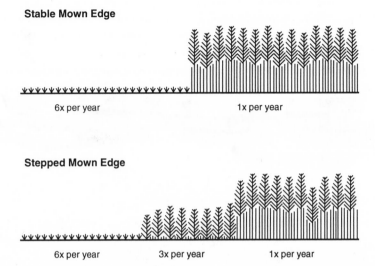

Stable Mown Edge

6x per year

1x per year

Stepped Mown Edge

6x per year

3x per year

1x per year

Figure 3 When the mown borders of wildflowers are moved in stages away from the meadow toward the surrounding turf, a more gradual, or stepped, edge results. This stepping produces a more visually attractive edge at the end of the season, and encourages the natural expansion of the meadow.

removes undesirable species. As a general maintenance practice it has obvious limitations due to concerns for safety and air quality. (See the chapter on the Indiana Dunes by Whitman et al.)

Conclusions

Wildflower meadows are sustainable landscapes. From ecology we have learned that biological communities rarely reach a permanent or stable equilibrium condition. Living systems are in a constant struggle to develop, to change, and to respond to disturbance. Control of this process is the fundamental goal of landscape maintenance, whether it be lawn mowing or ecological restoration. Sustainable landscapes achieve this control through informed management of natural processes rather than through heavy inputs of labor or chemicals. Sustainable landscapes achieve the necessary control in a nontoxic, cost- and energy-effective manner and promote a greater awareness of natural processes and ecological interrelationships between humans and their environment.

Wildflower meadows as sustainable landscapes have an important place in the urban environment. After all, if landscapes, i.e., areas without buildings or roads, do not contribute to biodiversity, where will it exist in the urban landscape?

These incremental statements may help to support the development of a sustainable aesthetic in landscapes that is based on ecological principles rather than on preconceptions of romanticized, idealized garden styles.

Complete acceptance of alternative landscape treatments like wildflower meadows will involve integration of additional factors into the prevailing aesthetic values. In sustainable terms, *beautiful* will signify that a landscape is nontoxic, not dependent on imported subsidies of agrochemicals for its existence, and that it will make a positive contribution to local biodiversity as a refuge or reservoir for native species of plants and animals.

References

Ahern, J. 1988. The establishment and maintenance of wildflowers and native woody plants in the highway landscape. *Proceedings: Fourth Symposium on Environmental Concerns in Right-of-Way Management,* ed. R. W. Byrnes, and H. Holt. West Lafayette, IN: Purdue University.

———. 1990. Preparation and establishment techniques for meadow gardens. *Meadows and Meadow Gardening* 5(1): 14–20. Framingham, MA: New England Wildflower Society.

Andersen, B. 1990. Using prescribed burns as a prairie management tool. *Wildflower* 3(2):27–33. Austin: National Wildflower Research Center.

Blake, E. L., Jr. 1990. Translating the beauty: Idealized nature in an urban context. *Wildflower* 3(1):14–23. Austin: National Wildflower Research Center.

Bones, J. 1975. Riotous flora by a Texas wayside. *Audubon.* 7(75):36–39.

Bos, H. J., and J. L. Mol. 1979. The Dutch example: Native planting in Holland. In *Nature in cities,* ed. I. C. Laurie. New York: John Wiley.

Burley, C., and J. B. Burley. 1991. Weedpatch syndrome II: an examination of the wildflower movement in North America. *Landscape Research* 16(1):41–45.

Diekelman, J., and R. Schuster. 1982. *Natural landscaping: Designing with native plant communities.* New York: McGraw-Hill.

Emerson, B. H., 1990. Meadow management. *Meadows and Meadow Gardening.* 5(1):29–32. Framingham, MA: New England Wildflower Society.

Evans, D. 1987. Take a ride on the wild side. Master's project, Department of Landscape Architecture and Regional Planning, University of Massachusetts at Amherst.

Fabos, J. G. 1985. *Land use planning: From global to local challenge.* New York: Chapman and Hall.

Gilbert, B. 1987. In from the fields, wildflowers find a new welcome among gardeners. *Smithsonian* 18(1): 37–45.

Gore, A. 1992. *Earth in the balance: Ecology and the human spirit.* New York: Houghton Mifflin.

Grese, R. E. 1990. Historical perspectives on designing with nature. In *Restoration 89: The new management challenge,* ed. H. G. Hughes and T. M. Bonnicksen. Proceedings of the First Annual Meeting of the

Society for Ecological Restoration, 16–20 January 1989, Oakland, CA. Madison, WI: Society for Ecological Restoration.

Hough, M. 1984. *City form and natural process.* New York: Van Nostrand Reinhold.

Kaplan, R. 1984. Dominant and variant values in environmental preference. In *Environmental preference and landscape maintenance: Proceedings of the symposium,* 21–23 October 1983. Connecticut College, New London.

Laurie, I. 1979. Urban commons. In *Nature in cities,* ed. I. C. Laurie. New York: John Wiley.

Leopold, L. B., and T. Dunne. 1978. *Water in environmental planning.* New York: W. H. Freeman.

Longland, D. R. 1985. Lawns, meadows or both? *Wildflower Notes* (1):11–18. Framingham, MA: New England Wildflower Society.

Martin, L. C. 1986. *The Wildflower Meadow Book.* Charlotte, NC: Fast and McMillan.

McCommon, M. 1983. A blooming boom in Texas, *National Wildlife,* Aug/Sept. 4–5.

Meyers, E. M., ed. 1990. *Meadows and meadow gardening* 5(1):25–28. Framingham, MA: New England Wildflower Society.

Minnesota Roadside Wildflower Task Force. 1988. Final report and recommendations. October.

National Wildflower Research Center. 1989. *Wildflower handbook.* Austin: Texas Monthly Press.

Pollan, M. 1991. Abolish the White House Lawn. *New York Times,* 5 May.

Radosevich, S. R., and J. S. Holt. 1984. *Weed ecology: Implications for vegetation management.* New York: John Wiley.

Spirn, A. W. 1984. *The granite garden.* New York: Basic Books.

World Commission on Environment and Development. 1987. *Our common future.* New York: Oxford University Press.

Common and Botanical Names of Species Mentioned in the Text

Common name	Botanical name
Canada goldenrod	Solidago canadensis
crabgrass	Digitaria sp.
cup plant	Silphium perfoliatum
dandelion	Taraxacum officinale
Joe-Pye weed	Eupatorium maculatum
lance-leaved coreopsis	Coreopsis lanceolata
New York ironweed	Vernonia noveboracensis
ox-eye daisy	Chrysanthemum leucanthemum
black-eyed Susan	Rudbeckia fulgida
spotted knapweed	Centaurea maculosa
swamp milkweed	Asclepias incarnata
Turk's-cap lily	Lilium superbum

The Indiana Dunes: Applications of

Landscape Ecology to Urban Park Management

Richard Whitman, Daniel B. Fagre, Noel B. Pavlovic,
and Kenneth L. Cole

Introduction

Established by Congress in 1966, the Indiana Dunes National Lakeshore is located on the southern shores of Lake Michigan near our nation's third-largest city, Chicago. Dominated by glacial moraines, lakes, ponds, bogs, streams, lagoons, pannes, beaches, dunes, and marshes ranging from two thousand to twelve thousand years in age, Indiana Dunes consists of about fifteen thousand acres of natural areas and provides outdoor recreational facilities for over two million visitors per year.

Two of the major purposes of establishing the national lakeshore were its historical significance as the birthplace of modern ecology in North America and its unique floristic diversity. This floristic diversity was one of the features that inspired Henry Cowles to study the dunes environment and to formulate his seminal ideas on plant succession (Cowles 1899), which ultimately had a major impact on ecological theory. The Indiana Dunes region was well known for its unusual juxtaposition of plant distributions as early as 1927 (Lyon 1927; Pepoon 1927; Peattie 1930; Deam 1940). Plants normally found in biomes distant from one another form rare and remarkable assemblages within the foredune communities, the biological assemblage occupying the beachward dune facies.

This phenomenon has arisen in part because of the diverse geological features concentrated within the boundaries of the national lakeshore. Of equal importance, the Indiana Dunes were in an advantageous geographic position to sustain biological diversity because floras there changed dramatically during various glacial ages. As a result of its diverse microhabitats and the moderating influence of Lake Michigan, the dunes region was able to retain remnant populations when surrounding conditions changed. In essence, this area serves as a refugium for plants at the extremes of their ranges.

Because of the diverse mosaic of climatic, edaphic (soil), and hydrologic microhabitats, the present Indiana Dunes National Lakeshore boundaries encompass the highest plant species diversity per acre of any park in the National Park system.

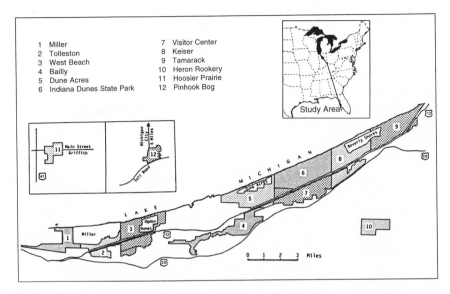

1 Miller
2 Tolleston
3 West Beach
4 Bailly
5 Dune Acres
6 Indiana Dunes State Park
7 Visitor Center
8 Keiser
9 Tamarack
10 Heron Rookery
11 Hoosier Prairie
12 Pinhook Bog

Figure 1 Indiana Dunes National Lake and designated park units

It is conceivable that the floristic diversity may have been increased by the east-west movements of both Indians and, later, white Europeans as they were forced to travel around the southern tip of Lake Michigan. Over fourteen hundred species are found in just fifteen thousand acres, making the Indiana Dunes National Lakeshore a prominent hub of biodiversity in the midst of the urbanized region of Chicago and northwest Indiana.

Threats to Resources

Perhaps not surprisingly, some of the same factors favoring the biodiversity in this area have also led to the current threats to that biodiversity. The Indiana Dunes area is a heavily used modern transportation corridor crossed by abundant highways and railroads. The disturbance of natural communities began over a hundred years ago with logging and altered fire frequencies. The heterogeneity of natural habitats was reduced with the advent of agriculture, the draining of wetlands, extensive settlement, and finally, industrialization.

The south shore of Lake Michigan is one of the nation's leading regions of steel production, chemical plants, and related industries (Mayer 1964). The establishment of the Indiana Dunes National Lakeshore protected, at the last minute, the remaining natural areas from complete development. Consequently, the configuration of Indiana Dunes was not planned to preserve an intact system, but only to safeguard the remaining natural "jewels." Initially, this strategy presented no threats to the integrity of the protected areas because surrounding intact natural landscapes acted as effective buffers against the effects of human-dominated environments. However, as development eliminated, and continues to eliminate, natural landscapes, the long-term sustainability of these protected areas becomes more precarious.

Much of the envelopment of the Indiana Dunes is linked to the urban expansion of Chicago, forty-five miles away. The expansion of urbanization into the dunes region, coupled with local increases in development, has functionally isolated the natural areas of Indiana Dunes from other regional natural areas. Encirclement by such developed areas creates an effective barrier to some species interchanges, which are further constrained by local isolation between some of the units at Indiana Dunes. Finally, habitat fragmentation, both natural and unnatural, within already small, isolated units creates small habitat patches that are inviable for the preservation of some of the native plant and animal species.

The trend toward isolation and fragmentation has numerous and often compounding effects on the viability of the protected areas as a system and on their biological value. For instance, the reduced ingress of individual animals or plant [seed] into the protected area jeopardizes population viability, the maintenance of genetic diversity, and the likelihood of recolonizing an area following some type of disturbance. The inability of animal populations to stabilize for lack of a means for surplus animals to disperse can lead to habitat damage. The reduced capability of populations in small, isolated habitats to endure random events such as storms and disease outbreaks means some species will become locally extinct by chance. Finally, the relatively higher ratio of outside threats, such as exotic species or pollution, to inside resources for small habitat patches indicates that a greater level of disturbance will prevail in these natural areas.

At issue is not only the sustainability of natural areas in urban settings but also the larger issue of managing, ultimately, to achieve sustainable cities. Humans have biological and psychological needs for naturalness, ranging from the aesthetic and spiritual to the intellectual and educational. The pursuit of these needs at the Indiana Dunes over the past century is examined by J. Ronald Engel (1983) in his book *Sacred Sands: The Struggle for Community in the Indiana Dunes*.

Naturalness also facilitates our physical survival on a global basis. Without viable, self-sustaining natural areas integrated into largely urban regions, advanced levels of public understanding concerning nature cannot be established or maintained. Such public awareness is increasingly necessary for making appropriate societal decisions to ensure human survival. Thus, these issues of sustaining natural systems in fragmented landscapes transcend the need of urban people for recreational outlets or aesthetic gratification on weekends.

Research Focus

The biological problems of relatively small and isolated natural areas embedded in heavily impacted landscapes provide fertile ground for research into the underlying relationships between the spatial characteristics of natural systems and their biological dynamics. In recent years these types of investigations have been rapidly developed in the field of landscape ecology (Turner and Gardner 1990). One of the major tasks in maintaining sustainable cities with integrated natural areas is maximizing the viability of remnant ecosystems that now receive many of their supporting functions from outside the protected areas. The problem is to reduce the vulnerability of natural areas when those supporting functions are threatened. How do we balance the need for the preservation of these functions with the competing need for infrastructure support of the city itself with its demands in industry, housing, and transportation? By maximizing the spatial patterns, landscape ecology potentially provides the capability to design a skeletal, yet intact, system using a minimum amount of land space.

Landscape theory suggests that a given amount of acreage of a specific vegetation type will support distinctly different assemblages and abundances of animals, depending on the distribution of that acreage into habitat patches of different numbers, sizes, shapes, and distances between the patches (Forman and Godron 1986). In other words, spatial patterns of natural landscapes are just as important as the total amount of natural landscape. At the Indiana Dunes National Lakeshore, potentially productive initial investigations should revolve around the size and shape of each protected area, and its distance from the others. The spatial analysis should be carried out further to examine the patterns of individual habitats within the protected areas and the impacts of these spatial attributes on populations.

The following paragraphs summarize salient efforts by the Indiana Dunes National Lakeshore Research Division to address scientific aspects of managing significant natural resources within an urban setting. Although the research program over the last twenty years has been diverse, the focus of this paper will be on current studies and problems as they relate to the ecological fragmentation of the park, with emphasis on plant, water, and animal resources.

Impacts of Ecological Fragmentation on Plant Resources

The Indiana Dunes National Lakeshore is charged by Congress with preserving the floristic diversity of the dunes. A total of 1,134 native vascular plants (and several hundred exotic plants) are found in the National Lakeshore; 14 percent of the native plants have been discovered in the last decade (Wilhelm 1990). No plant species are endemic to the Indiana Dunes region alone. However, of the known vascular species, 94 are listed by Indiana as state-endangered and state-threatened. The U.S. Fish and Wildlife Service lists 1 threatened plant on the federal list, Pitcher's thistle (*Cirsium pitcheri*), and 5 candidates for listing: fame flower (*Talinum rugospermum*), Hall's sedge (*Scirpus hallii*), bog bluegrass (*Poa paludigena*), dune goldenrod (*Solidago spathulata gillmannii*), and aromatic sumac (*Rhus aromatica*). Twelve species are reliably known to have been extirpated thus far.

Described below are some of the external and internal threats that have directly or indirectly affected rare plant species at the Indiana Dunes National Lakeshore. These examples will serve to illustrate that biological preservation in a highly fragmented and altered urban landscape requires active monitoring, management, and vigilance.

Many plant species have been impacted by shoreline erosion resulting from high lake levels. The erosion is exacerbated by human-made structures that alter lateral sand flow (School of Civil Engineering 1986). Sand bluff erosion in 1986–87 reduced the only Indiana rice grass (*Oryzopsis asperifolia*) population from eleven to two plants when the bluff collapsed into the lake (Pavlovic and Bowles, in review). The two surviving plants were cross-pollinated and twelve seeds were produced in an attempt to reestablish the population. Subsequently, one plant died, but one seed was viable and produced a new plant. Before any new plants were transplanted in the field, a new large population was discovered in the Dunes State Park, in 1988. Plans are underway to restore the population, using seed from the Dunes State Park. Several sites will be utilized, to reduce the risk of local extirpation through recurrent shoreline erosion, but a sustained monitoring effort will be required to maintain this species at Indiana Dunes.

Trampling by humans in combination with shoreline erosion has had a negative impact on some species. The loss of foredune habitats through shoreline erosion has eliminated the federally threatened Pitcher's thistle from the foredunes and confined this species to isolated habitats where wind has excavated open sand (i.e., dune blowouts) (Cowles 1899; Pepoon 1927; and Peattie 1930). At the park's West Beach area, the local population of Pitcher's thistle are smaller, fewer in number, and more scattered than on adjacent dunes where visitor access is controlled (McEachern, pers. comm.). This suggests that human trampling has a negative effect on Pitcher's thistle population. Circumstantial evidence also suggests that visitor use on the beaches lowers the population levels of, and in some cases eliminates, the annuals such as the state-threatened sea rocket (*Cakile edentula*) and sand spurge (*Euphorbia polygonifolia*) that are typical of the shoreline (Bowles et al. 1986a).

Pitcher's thistle also appears to need a spatial and temporal mosaic of bare sand to colonize as part of its life history. The loss of foredune habitats may have fragmented Pitcher's thistle dispersal corridors to other habitat mosaics suitable for colonization beyond a tolerable level, and may account for its reduction along the shores of Lake Michigan. The landscape ecology research of Kathryn McEachern will assist the National Park Service (NPS) in preserving viable metapopulations of this unique sand dune herb by identifying favorable spatial configurations of habitat (McEachern et al. 1989).

The more stabilized inland dunes are dominated by black oak savanna and prairie communities. Previously, these communities typically experienced frequent low-intensity fires. However, both increasing landscape fragmentation and fire suppression efforts over approximately forty years have nearly eliminated such fires. The result is that some rare species were extirpated from their natural habitat by the buildup of litter, the closing of the tree canopy, and the infrequent but intense fires caused by heavier fuel loads. Many of these species have invaded adjacent anthropogenic habitats as the landscapes changed (Bowles et al. 1990). Some species such as fame flower, false heather (*Hudsonia tomentosa*), and sea beach needle grass (*Aristida tuberculosa*) now survive along trails and roadsides and in sand-mined areas. The effectiveness of restoring fire frequencies to recreate previous habitats for these plants will depend to some degree on the spatial arrangement of the existing populations in the altered habitat mosaic.

The ecological research by Noel Pavlovic is designed to examine the spatial needs of fame flower and should assist the National Park Service in restoring this succulent herb to its natural dune habitat: small- to medium-sized gaps where plant competitors are absent. These gaps are maintained by fires that remove dead plant litter, by the activity of moles churning the soil, and perhaps by other disturbances by small animals such as diggings. It is paradoxical that fame flower grows in savannas and prairies where fires are frequent, because the plants are fire-sensitive and the species lacks an appreciable between-year seed bank. The paradox is explained by three facts: (1) sensitivity to fire damage declines as plant size increases; (2) the larger the disturbance patch, the greater the probability of escaping high fire temperatures; and (3) fire intensity decreases with increasing frequency. The restoration of fame flower will require the restoration of the plant community, the reintroduction of fire and disturbance processes to maintain small patches of bare soil, and the design of landscapes to ensure adequate dispersion of fame flower populations.

Another plant impacted by habitat fragmentation of the dunes region is false heather. The small easternmost population of false heather has declined in its blowout habitat. In 1981, black oaks surrounding this tiny blowout were trimmed to reduce shading, which was thought to be contributing to poor vigor in the plants. In this case, however, a proximal cause for its decline appears to be deer trampling since a major deer trail traverses the habitat. Deer populations are moderately high at the Indiana Dunes in the absence of any top carnivores and of hunting by humans, and because of a favorable habitat mosaic offering ample food and cover. In 1989 five plants remained and an additional nine plants were re-introduced to this site (Plampin 1989). These plants were collected from private property adjacent to the park. By 1991 (Plampin 1991), none of the original plants and only four of the reintroductions were alive. Successful restoration and man-agement of this population will be closely tied to the overall management of habitat mosaics as it influences deer populations. Spatial arrangements must be sought that will reduce the need for direct intervention to protect the plants from deer damage.

Finally, ditching, haying, industrial activity, and cranberry farming caused fur-ther habitat fragmentation for plants by altering the hydrology of the Great Marsh (Cook and Jackson 1978). The Great Marsh was the continuous wetland complex that extended from Gary to Michigan City, Indiana, between the Calumet and Tolleston/Nipissing dune ridges. The central heart of the Great Marsh was de-stroyed in the 1960s to construct the Bethlehem Steel Mill, causing one of the larger changes in hydrology and habitat continuity. Consequently, the sedge meadows that formerly surrounded Cowles Bog (a graminoid fen) in the Great Marsh are now dominated by cattails (Wilcox et al. 1984). Altered hydrology and the absence of fire in the grassy fen have reduced showy lady's-slipper (*Cypripedium reginae*) and white lady's-slipper (*Cypripedium candidum*) respectively to one and several small clumps (Bowles 1986b, 1988, 1989). Prescribed fires and elimination of the industrial ponds, in conjunction with cattail removal and sedge meadow restora-tion, will rejuvenate these and other graminoid fen species at Cowles Bog. How-ever, long-term sustainability will require management of the spatial mosaic to maintain metapopulations of these restricted species.

Historical Perspectives on Ecological Fragmentation

The wide assortment of substrate types and water regimes at the Indian Dunes combines to create a complex habitat mosaic embracing plants typical of the boreal forest, the Atlantic coastal plain, the eastern deciduous forest, and the Great Plains. Because most of the natural habitats are continually undergoing rapid change, knowledge of the vegetation history is particularly relevant for effectively maintaining both the vegetative communities and the restricted plant species. This vegetation history can provide a benchmark against which the effectiveness of current management and restoration strategies may be compared.

The development of housing and farming in areas of former prairie has caused the prairie to be one of the most restricted dune communities. Furthermore, the remaining prairie is quickly shrinking. Paleoecological and historical techniques were used to examine the history of a prairie area at Indiana Dunes in order to understand why it was shrinking so rapidly in modern times (Cole et al. 1990). The analysis of fossil pollen from a sediment core revealed that although the prairie had undergone a slow transition from pine forest to prairie/oak savanna over several thousands of years, the rate of change during the last 100 years had been greatly accelerated after settlement (Cole 1988). Furthermore, fire history based upon fire scars on trees and stand structure demonstrates that although regular fires continued to burn the prairie until about 1960, no recent fires have occurred, because of fire exclusion by humans (Taylor 1990). These findings underscore the critical role played by fire in maintaining prairie habitats. As the remaining prairies are virtual islands in the surrounding forests, a landscape ecology perspective will need to be brought to bear to restore these prairie remnants so that they function as linked communities.

The dunes have been the focus of several classical plant ecology studies defining and examining the concept of primary plant succession (Cowles 1899; Cowles 1901; Peattie 1930; Olson 1958). More recently, the classical successional sequences found on the dunes have been investigated by means of paleoecological methods. These techniques allow the reconstruction of actual long-term successional sequences at a specific location on the basis of fossil data, whereas earlier researchers simply inferred sequences based on scattered dunes of differing ages. Studies such as those by Richard Futyma (1985), Steve Jackson (1988), and Kenneth Cole (1990) and their colleagues demonstrate changes in the dunes over the last four thousand years that parallel those inferred by Henry Cowles.

The composite picture from these investigations is of a dunes landscape that was periodically burned until Anglo-European settlement. Soon after, steam locomotives began to travel through the area, dispersing sparks and increasing the frequency of fire. Subsequently, many wetlands were drained, opening up expanses of new habitat to intense fire. As diesel locomotives replaced steam ones and fire control efforts became effective, virtually all fires ceased.

Fire Management and Plant Resources

Native prairie vegetation is clearly adapted to flourish under periodic fire. As the historical record indicates, in the absence of fire woody species invade and suppress the prairie with shade. The management of fire-adapted vegetation in close proximity to urban areas creates special difficulties for managers. At the Indiana Dunes National Lakeshore, remnant patches of rare prairie species occur adjacent to expensive residences and even petroleum tank farms. Burning these prairies requires special precautions, well-trained fire teams, and carefully prepared contingency plans.

The fires used in managing these areas are known as "prescribed burns." They are "prescribed" (by a scientist or manager) as being necessary for the well-being of the patient (the prairie habitat). Some justification for prescribed burning comes not from the need to restore prairies but from the realization that in fire-prone areas, woody fuels from plants will continue to collect until a fire occurs. If there have been no fires in a lengthy period of years or decades, then the large amount of fuel on the ground can result in an intense, uncontrollable wildfire. In contrast, regular prescribed fires burn at a low intensity, relieve the buildup of woody fuel, and reduce the wildfire risk for urban areas adjoining natural habitats. They can be carefully planned to occur during a safe climatic burning "window," and when fire crews can be fully deployed at the location.

The essential elements of a prescribed burning program are (1) an understanding of the effects of fire on the habitat; (2) a well-trained fire crew that will make sure the fire follows the plan; and (3) a fire plan outlining the conditions under which a fire will be permitted. Such a prescribed burning program has been implemented at Indiana Dunes and followed by intensive monitoring of the results on prairie vegetation.

Armed with our rare plant inventory, monitoring programs, and judicious research efforts, we should be able to preserve and maintain biodiversity and the rare species for which the Indiana Dunes are famed. By examining the spatial patterns of existing habitats and by managing communities rather than individual species, we hope to restore and preserve the natural ecosystem processes that ensure the existence of all native species. Ultimately, this is a more rational and effective goal for the management of rare plant species.

Impacts of Ecological Fragmentation on Water Resources

Water played a significant role in the geological, ecological, and cultural history of northwest Indiana. Water in the form of ice carved Lake Michigan, and the lake's receding shoreline formed the ridge-and-swale topography characteristic of the dunes (Thompson 1987). Between these ridges there formed a linear series of interdunal wetlands that parallel the Lake Michigan shoreline. These interdunal ponds are chronologically older and successionally more advanced the further landward they lie. At the turn of the century, Henry Cowles and his student, Victor Shelford, recognized that this sequence of progressively older wetlands presented an ideal situation for describing their theory of aquatic succession. Shelford's studies laid the foundation for modern theories of natural eutrophication (Shelford 1912); unfortunately, most of Shelford's original ponds have fallen victim to urbanization.

The interdunal wetlands have been extensively fragmented through industrial, residential, agricultural, roadway, and railroad development entailing diversions, filling, dredging, and draining. In spite of these disturbances, wetlands still cover large areas of the Indiana Dunes National Lakeshore and are potent factors in the ecology of the dunes landscape. These aquatic habitats include an almost complete series of ponds in the Miller Woods portion of the park, the large interdunal pond of Long Lake, the nationally recognized Cowles Bog, and the Great Marsh. Although the more extensive eastern portion of the Great Marsh has been greatly reduced in depth and area by dredging, the marsh is one of the most significant landscape features of the park. The principal ditches or modified streams that used to drain the Great Marsh include the Derby and Kintzele ditches, and Dunes Creek.

A number of problems have arisen from the fragmentation of water resources. The first is the impact on terrestrial habitats. Any restoration or mitigation efforts to change the spatial distribution of vegetative communities in order to better provide a skeletal ecosystem must necessarily be tied to the management of water resources.

These water resources are problematic in their own right. Normally one might view water coursing through a national park as an important aesthetic and natural resource. The ditches and artificial streams at Indiana Dunes, however, tend to be major concerns for park management because they are iron-rich. As they reach Lake Michigan, they naturally introduce detritus, tannins, and precipitated iron oxide, which bathers find unsightly against the blue waters of the lake.

In addition to substances that have an aesthetic impact, these ditches introduce coliform bacteria into the lake. On occasion, the bacterial levels exceed state regulations for recreational swimming and bathing. Studies have demonstrated that the bacterial sources are nonpoint in origin and are probably derived from animals that naturally occur in riparian and wetland communities. Nevertheless, the state and federal regulations do not differentiate between animal- and human-derived bacteria, and the beach often must be closed to swimming during peak visitation in the summer. The obvious solution to this immediate problem, and to the ultimate issue of water resource fragmentation, would be to fill in these drainages and restore the Great Marsh to its original expanse. Unfortunately, political realities do not allow for the inundation of existing residential and industrial developments, and no other feasible solutions are presently apparent.

Just to the west of the "industrial island" that delineates the western extent of the

Great Marsh is the Burns Harbor waterway. The waterway (completed in 1926) connects Lake Michigan to the Little Calumet River, which once continued to flow westward and supported about twenty thousand acres of wetlands (Moore 1959). The dredging of the Little Calumet and the development of the harbor diverted the eastern portion of the Little Calumet River directly into Lake Michigan and effectively reversed the flow of the western portion of the river. The channelized Little Calumet River drained municipal sewers, urban areas, and farmland into the harbor. The modified flow dynamics also increased bank erosion and damage to riparian residential property. The harbor outlet now introduces clays/silts, coliform bacteria, sewage waste, high nutrients, and organic and inorganic contaminants into Lake Michigan. The nearby West Beach is one of the most popular beaches in the Great Lakes region, hosting approximately 225,000 visitors per year. As the contaminants may drift toward the West Beach area, the modifications to the Little Calumet continue to provide significant concerns for natural resource management at Indiana Dunes.

Other types of ecological fragmentation of the lakeshore are not as obvious as the preceding examples. Just above the confluence of the Salt Creek and Little Calumet rivers, temperature differentials between the river water and Bethlehem Steel's thermal effluent effectively block the migratory patterns of salmonids as they move upstream in the spring and fall. It is suspected that migrating salmon elect to move into Salt Creek rather than attempt to pass through the thermal boundaries of this effluent. A number of solutions have been proposed; they include importing Lake Michigan water to cool the effluent, establishing cooling ponds, and constructing fish passageways. To date, no mitigation approaches have been implemented.

Impacts of Ecological Fragmentation on Animal Resources

A series of linked research efforts have been initiated to bring a landscape focus to the study of the distribution and abundance of animals in relation to existing habitats. The Indiana Dunes' geographic information system (GIS) is being used to analyze the distribution and spatial attributes of all habitat patches in the foredune community. A data base of patch attributes will be created, and animal distribution and abundance patterns will be quantitatively compared against the data base to elucidate the degree to which spatial attributes of habitats constrain animal populations. The landscape will be characterized in terms of heterogeneity, potential for contagion, connectivity, and other measures at different scales. This spatial analysis will provide an alternative hypothesis for the larger, program-wide null hypothesis that animal abundance and distribution are directly related to the total extent of appropriate habitat and do not vary significantly as a function of landscape characteristics such as patch size, shape, and connectivity.

This research program is in its early stages as of this writing, but traps have been monitored to characterize the mammal diversity and abundance in four habitats of the foredune community. Replicate trap grids are located in two patches of each habitat type. These grids are arranged in a counterbalanced design in a transect from the western part of the park to the eastern (a distance of fifteen miles). Grids are operated for five or more nights during the spring and early summer, and again in the late summer and early fall. Population estimates based on marking and recapture are made possible by ear tags for the small mammals. From the results of these efforts, predictions will be made as to species diversity, richness, abundance, and commonness by habitat type, and as to abundance and commonness by animal species. Follow-up tests will focus on one habitat type and one species with larger sample sizes to further characterize the impacts of the spatial attributes of habitat patches on mammal distributions.

Concurrently, captured small mammals are being checked for the presence of the deer tick, *Ixodes dammini,* to assess the potential spatial distribution of animals carrying Lyme disease at Indiana Dunes. From these data, a "landscape" of risk to park visitors can be generated with GIS technologies, placing previously unavailable tools for decision making in the hands of park managers.

In a manner similar to that of the small mammal studies but using different techniques, white-tailed deer distributions are being characterized for various habitats. A GIS model that will predict deer density by habitat patch will be the final product of an analysis of deer population data in relation to the spatial characteristics of Indiana Dunes' landscapes. This, too, will be an iterative process whereby further tests will be made in which individual deer will wear radio transmitters to determine when and how they choose to utilize various patches.

From the models of animal distributions refined by landscape theory, managers will have powerful decision-making tools at their disposal. By utilizing a geographic information system, what-if scenarios can be quickly generated to assess potential impacts of development or controlled fires on species distribution and abundance. Managers can identify areas of their landscape where it is critical to expand habitats, areas where, for example, a relatively small increase in acreage may significantly increase the biological value of that site to a specific species because the site now has the minimum space necessary for a territory for that animal. Key dispersal corridors can be identified with the landscape approach and protected to ensure connectivity and animal movements between habitat patches. Other corridors can be artificially reestablished to strengthen the ecosystem integrity with minimal impacts on land uses between habitat patches.

It is even possible to use landscape theory to manage overpopulation problems or conflicts with humans. Deer travel corridors could be severed with relatively little habitat change to greatly reduce their damage to crops or landscaping, or to direct them away from curves in roads where deer-vehicle collisions are common.

Conclusion

Whatever the specific applications, the use of landscape principles to analyze spatial patterns in parks is likely to provide us with the ability to design and maintain skeletal ecosystems of maximum efficiency and to mitigate effectively the impacts of ecological fragmentation. This ability to understand the spatial aspects of landscapes and to design solutions to problems makes us capable of more skillful land management with respect to protecting the plant, water, and animal resources of the Indiana Dunes National Lakeshore. More broadly, this ability will help to achieve the highest possible biological value and ecosystem vitality for our urban parks with the least space, and thus help to integrate them successfully into the wider sustainable urban landscape.

Acknowledgments

We gratefully acknowledge the assistance of Kevin Kennedy for providing graphical preparation and Diane M. Daum for her editorial services.

References

Bowles, M. L., W. J. Hess, and M. M. DeMauro. 1985. An assessment of the monitoring program for special floristic elements at the Indiana Dunes National Lakeshore: Phase 1. The endangered species. Unpublished report.

———. 1986a. An assessment of the monitoring program for special floristic elements at the Indiana Dunes National Lakeshore: Phase 11. The threatened species. Unpublished report.

Bowles, M. L., W. J. Hess, M. M. DeMauro and R. Hiebert. 1986b. Endangered plant inventory and monitoring strategies at Indiana Dunes National Lakeshore. *Natural Areas Journal* 6(1): 18–26.

Bowles, M. L., M. M. DeMauro, N. Pavlovic, and R. Hiebert. 1990. Effects of anthropogenic disturbances on endangered and threatened plants at the Indiana Dunes National Lakeshore. *Natural Areas Journal* 10:187–200.

Bowles, M. L. 1988. A report on special floristic elements at the Indiana Dunes National Lakeshore: New species monitoring and update of selected existing populations. Unpublished report.

———. 1989. A status report on endangered and threatened plants of the Indiana Dunes National Lakeshore: Monitoring of species new to the lakeshore and remonitoring of selected species. Unpublished report.

———. 1990. Report on the status of endangered and threatened plants of the Indiana Dunes National Lakeshore: Monitoring of species new to the lakeshore and remonitoring of selected species. Unpublished report.

———. 1991. Report on the status of endangered and threatened plants of the Indiana Dunes National Lakeshore: Monitoring of species new to the lakeshore and remonitoring of selected species. Unpublished report.

Cole, K. L., R. S. Taylor, and K. F. Klick. 1990. The history of an Indiana sand prairie: The effects of time, water, and fire. In-house report to Indiana Dunes National Lakeshore. (Submitted for publication.)

Cole, K. L. 1988. Historical impacts on communities in disequilibrium. In Symposium on plant succession. Proceeding of the First Indiana Dunes Research Conference, ed. K. L. Cole, R. D. Hiebert, and J. D. Wood. Gary, IN.

Cook, S. G., and R. S. Jackson. 1978. *The Bailly Area of Porter County, Indiana: The final report of a geohistorical study undertaken on behalf of the Indiana Dunes National Lakeshore.* National Park Service, Porter, IN.

Cowles, H. C. 1899. The ecological relations of the vegetation on the sand dunes of Lake Michigan. *Botanical Gazette* 27:95–117; 167–202; 281–308; 361–91.

———. 1901. Physiographic ecology of Chicago and vicinity: a study of the origin, development, and classification of plant societies. *Botanical Gazette.* 31:73–108.

Deam, C. C. 1940. *Flora of Indiana.* Indianapolis: Indiana Department of Conservation.

Engel, J. R. 1983. *Sacred sands: The struggle for community in the Indiana Dunes.* Middletown, CT: Wesleyan University Press.

Forman, R. T. T., and M. Godron. 1986. *Landscape ecology.* New York: John Wiley.

Futyma, R. P. 1985. *Paleobotanical studies at Indiana Dunes National Lakeshore.* National Park Service, Porter, IN.

Indiana Dunes National Lakeshore. 1989. *Resource management plan and environmental assessment.* National Park Service, Porter, IN.

Jackson, S. T., R. P. Futyma, and D. A. Wilcox. 1988. A paleoecological test of a classical hydrosere in the Lake Michigan dunes. *Ecology* 69:928–36.

Lyon, M. W., Jr. 1927. List of flowering plants and ferns in the Dunes State Park and vicinity, Porter County, Indiana. *The American Midland Naturalist* 10:245–95.

———. 1930. List of flowering plants and ferns in the Dunes State Park and vicinity, Porter County, Indiana: supplement. *The American Midland Naturalist* 12:33–43.

Mayer, H. M. 1964. Politics and land use: The Indiana shoreline of Lake Michigan. *Annals of the Association of American Geographers* 54(Dec.): 508–23.

McEachern, K., J. A. Magnuson, and N. B. Pavlovic. 1989. *Preliminary results of a study to monitor Cirsium pitcheri in Great Lakes National Lakeshores.* Science Division, Indiana Dunes National Lakeshore, National Park Service, Porter, IN.

Moore, P. A. 1959. *The Calumet region: Indiana's last frontier.* Indianapolis: Indiana Historical Bureau.

Nelson, S. D., L. C. Bliss, and J. M. Mayo. 1986. Nitrogen fixation in relation to Hudsonia tomentosa—a pioneer species in sand dunes, northeastern Alberta. *Canadian Journal of Botany* 64:495–2501.

Olson, J. 1958. Rates of succession and soil changes on southern Lake Michigan sand dunes. *Botanical Gazette* 119:125–70.

Peattie, D. C. 1930. *Flora of the Indiana Dunes.* Chicago: Field Museum of Natural History.

Pepoon, H. S. 1927. *An annotated flora of the Chicago area.* Chicago: Chicago Academy of Sciences.

Plampin, B. 1989. Letter to Noel B. Pavlovic and Gerould Wilhelm. 13 December.

———. 1991. Letter to Noel B. Pavlovic. 10 June.

School of Civil Engineering. 1986. *Indiana Dunes National Lakeshore shoreline situation report.* West Lafayette, IN: Great Lakes Coastal Research Laboratory, Purdue University.

Shelford, V. E. 1911. Ecological succession. Part I. Stream fishes and the method of physiographic analysis. *Biological Bulletin* 21:9–34.

Taylor, R. 1990. Reconstruction of twentieth century fire histories in black oak savannas of the Indiana Dunes National Lakeshore. Master's thesis, University of Wisconsin-Madison.

Thompson, T. 1987. Sedimentology, internal architecture and depositional history of the Indiana Dunes National Lakeshore and State Park. Doctoral dissertation, Indiana University.

Turner, M. G. and R. H. Gardner, eds. 1990. *Quantitative methods in landscape ecology.* New York: Springer-Verlag.

Wilcox, D. A., S. I. Apfelbaum, and R. D. Hiebert. 1984. Cattail invasion of sedge meadows following hydrologic disturbance in the Cowles Bog wetland complex, Indiana Dunes National Lakeshore. *Wetlands* 4: 115–28.

Wilhelm, G. 1990. *Special vegetation of the Indiana Dunes National Lakeshore.* Report 90–02. National Park Service Midwest Region, Porter, IN.

Rethinking the Urban Park: Rediscovering Urban Solutions

Annaliese Bischoff

Introduction

Forest Park is a renowned 735-acre public park in Springfield, Massachusetts, an old manufacturing city on the Connecticut River, ninety miles west of Boston. For more than 100 years, Forest Park has served as an important natural asset to the city of Springfield as well as to the western Massachusetts region. Over time the role of the park and its relationship to nature have shifted and evolved to meet the changing needs of its users. The author will highlight the changing roles, reflecting five eras of the park's history in this case study. There are two objectives here. The first is to promote recognition of the diverse roles an urban park can play. The second is to inspire new thinking about current urban challenges and about the potential roles an urban park might play in helping to address them. The responses and solutions from earlier eras offer lessons today in rethinking city problems.

As a natural resource, Forest Park has always been revered for its great natural beauty. The striking landforms in the park are the result of ancient glacial carvings and postglacial erosion. Forest Park is full of rich, dramatic landscape, characterized by terrace escarpments, steep wooded ravines, and flat upland areas all radiating like fingers off a long central valley through which the Pecousic Brook meanders. There are over eighty different species of native plants with a dominance of mixed hardwoods (oaks, maples, and ashes) and softwoods (white pine and Canadian hemlock). Other native species include sycamore, beech, cedar, chestnut, red and pitch pine, and hornbeam. The rich, sandy loam soils continue to support an abundant, diverse, and healthy collection of plant materials. The landforms, brook, and native plants were particular features of Forest Park's original lands.

Pleasure Grounds: 1884–1900

In the first chapter of Forest Park's history, its primary role was to provide pleasure from the simplicity of its beautiful woodland character. In its first years, Forest

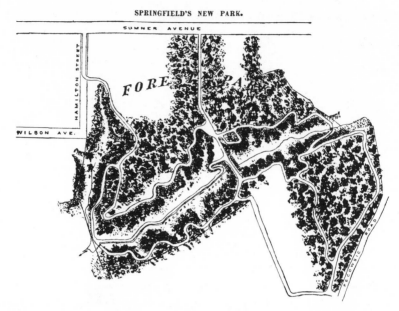

Figure 1   First known map of Forest Park, appearing in the *Springfield Sunday Republican*, 29 November 1885

Park was a source of passive recreation, mostly for the upper class. In 1884, Orick Greenleaf, a public-spirited paper manufacturer, began Forest Park with a sixty-five-acre land donation to the city of Springfield (fig. 1). Earlier, on 2 May 1871, Samuel Bowles, noted editor of the *Springfield Republican*, and Orick Greenleaf consulted with Frederick Law Olmsted, Sr., about a parkway scheme for the whole city. Additionally, Greenleaf met with Olmsted during his 2 May visit about a private lot Greenleaf wanted to purchase for his personal residence (Marston 1986). Beyond a newspaper article entitled, "Springfield and How to Improve It," in the *Springfield Daily Republican*, 6 June 1871, there was little further mention of a parkway scheme. However, Olmsted no doubt had influenced and perhaps inspired Greenleaf's donation.

The original purpose of Forest Park was to preserve a wild and natural landscape for the enjoyment of the public, as the name "Forest Park" implies. The mayor appointed a five-member park commission to direct its management. An article from the *Springfield Sunday Republican* describes the preservation intentions of the commission.

> Beyond the main ravine where the brook runs, lies perhaps a third of the park. Perhaps 10 acres on this side of the ravine are open grounds and here

the commissioners will probably carry out their idea of setting up picnic grounds for the convenience of people who wish to use the park for a day's outing. The trees are large and the shade heavy and cool, and the edges of the open land afford an opportunity for rustic seats and woodland walks without breaking up the forest effect of the whole, which it is the wise aim of the commissioners to preserve. . . . The object of the commissioners has not been to make a work of art, but to preserve so far as possible the wild aspect of the spot, which is one of the most beautiful in the region. (19 July 1885)

The commissioners hired the local private firm of Sackett and Reynolds to lay out carriage roads and footpaths. The firm built bridges, grottos, and springs, using local stone, rustic structures made with native hewn cedar, and small ponds impounding brooks.

Because access to this park in the remote southern end of the city was most available to those who could afford private carriages, many lamented the limited access. "It is a good half mile from the street car line to the gates and too long a walk for many of the people that most need and would best appreciate its cooling shade. Sometime doubtless the street cars will be extended to the park gates, but that is a matter of considerable expense that the street railway company cannot be asked to bear until there is a reasonable prospect of a return for the investment" (*Springfield Sunday Republican,* 19 July 1885). In 1890, the street railway company established a line of electric cars to the park entrance. About ten thousand visitors started traveling to the park each month. The middle class could easily afford the six-cent fare. By 1891, several substantial gifts of new land had enlarged the park to 330 acres. New carriage roads, new ponds, and new structures were built.

During this period, the park was functioning predominantly as a grand pleasure ground affording numerous opportunities for passive recreation in a natural woodland setting. The Sunday family picnic in Forest Park became a virtual institution for middle-class families. Throughout the 1890s, the park's range of facilities and amusements broadened. In 1893 the first baseball diamonds were added. The park was beginning to offer more active recreation opportunities. New lands donated in 1890, 1892, and 1894 by the world-renowned skate manufacturer, Everett Barney, contributed a new ornamental character to the Forest Park landscape. Barney, an avid horticulturalist with a passion for exotic ornamentals, owned award-winning lotus and lily collections. By 1900, in the 461 acres of Forest Park there were 320 plant species reported, only 80 of which were native. By virtue of both the original woodland character of the site and the subsequent ornamental landscapes, Springfield by the turn of the century had already gained much pride and benefit from its grand pleasure park (fig. 2).

Figure 2   Postcard of a spring in Forest Park depicting its woodland character.

Social Good: 1901–16

Financial stringency marked the year of 1901, and city park appropriations were cut by 20 percent. The era of work by private contracting came to an abrupt close with the creation of the post of park superintendent, first held by Carl E. Ladd. He instituted a labor force of city employees headed by civil service foremen.

The demands for more active and specialized recreation grew. Tennis, bicycling, football, basketball, and baseball were increasing in popularity, which the park accommodated. During the same period a growing number of prominent citizens wanted to give the park memorials, often in appreciation for their enjoyment of the park itself. The park responded to the needs of the new users and welcomed the memorial gifts.

Meanwhile, iron footbridges, cement dams, and concrete bridges were built in the park. The park department annals reported that the rustic cedar structures were abandoned because of their "short life span." More permanent materials began to replace the more natural, local ones.

By 1912 there was a new zoo that housed exotic monkeys and African lions. The earlier collection featured more "ordinary" species such as those in the extensive bird exhibit and the deer preserve. In 1914, an extensive rose garden was built and proved to be very popular in the city. In 1916, when Everett Barney died, he left the park with a substantial trust fund to tend the elaborate plant collections. The park was no longer simply a remote woodland spot serving as a scenic retreat for those who could afford to travel there. During this period Forest Park had broadened its programs to include more specialized activities emphasizing active, organized recreation. While whole families often stayed together during visits in the earlier era, users now were typically divided by age and gender into different recreation programs. Active recreation was considered to be for the social good of its users. The natural setting offered a supportive backdrop to the activities. The more structured administration of the park enabled a smooth operation of the increasingly diverse facilities designed to meet the needs of a growing Springfield population.

Public Service: 1917–40

The First World War interrupted the routine recreational programs of Forest Park. But it also inspired a response to changing needs. Starting in 1917, there was widespread use of the parkland for public vegetable gardens. A demonstration garden was constructed to help novice gardeners improve their produce. The public leased over five hundred garden plots yielding over six thousand bushels of potatoes alone in one season in Forest Park. Cabbage, beets, and turnips were raised, with an estimated value of $35,000. In 1918, in response to an influenza epidemic, officials swiftly turned the park into an outdoor hospital, nicknamed "Tent City." Forest Park was adapted quickly to meet significant, pressing needs of its time in an ever-evolving role of public service.

Under the influence of city forester Fletcher Prouty, Forest Park became increasingly important for its natural resources. In 1920, a sawmill was constructed to mill salvaged chestnut logs into lumber for use by city departments. The relationship of Forest Park to the rest of the city as the largest public open space in Springfield is illustrated in figure 3. In 1922 wood that was cut and cleared from the park was given to poor families for heating purposes. An active reforestation program was implemented in the park that same year. Over ten thousand trees, mostly white pines, were planted. The banks of the newly constructed thirty-five-acre Porter Lake were planted with over five thousand evergreens. The tree nursery at Forest Park supplied rhododendrons, laurels, and andromedas by the thousands for plantings on stream banks and roadsides. The program was well organized and highly productive.

In 1928, a summer camp for underprivileged children opened in the park. The success of the camp was reported as measured by the 1.6-pound average weight gain per child. Recreational activities offered for the general public included "open" dancing, "married" dancing, skating, and marble shooting.

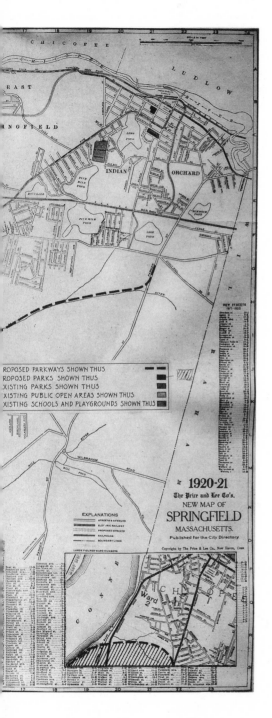

Figure 3   A map of the public open space in the city of Springfield, from 1921, from *A City Plan for Springfield, Mass* (1924)

During the Great Depression, Theodore Geisel (the father of "Dr. Seuss") served as the superintendent of parks. Thanks to abundant national work relief programs, the park was rich in labor for such improvements as nature trails and birdhouses. Although capital improvements were sparse, an old skating house from 1907 was converted into a "Trailside Museum" and served as a nature interpretation center.

During this time Forest Park assumed a strong role of public service in which the natural resources of the park were pragmatically managed and utilized. The relationship to nature was now rooted in productive management, not just admiration.

Enlarged Facilities: 1941–65

Between 1941 and 1965, the park greatly expanded its facilities. To serve the increasing population of the postwar period, the size and scale of the new facilities also increased. Concern for convenience as well as for safety was reflected in the more modern structures built in this era. Forest Park was now a destination for the new motoring public. Driving had fast become the new recreational pastime for the American family, which often traveled together on outings. The park landscape was newly rediscovered from the car window as a source of passive entertainment.

There was also an increasing passivity in the role of visitors, who became spectators in the park, especially at sporting events. In 1941 plans were completed for a new park grandstand, the Walker Memorial Grandstand. The city hoped it would be grand enough to attract a professional baseball team. Opening in 1949, it never did lure a pro team to town. The old, outdated concession was razed so that a newer, larger one could be built, while at the same time new parking lots were sited. For a time the need to retrofit the park for the onslaught of automobiles took attention away from nature.

The impact of the automobile went beyond parking lots and vehicular drives in Forest Park. The state proposed plans for Interstate 91 including an exit ramp that would jeopardize the historic Pecousic Villa in Forest Park. Pecousic Villa was the former Barney estate, which had been functioning as a museum for industrial exhibits since 1943. The beautiful structure along with its commanding views to the Connecticut River was sacrificed in 1959. By 1961 the park had lost about twenty acres of land.

Since 1961 the park has remained stable at 735 acres. During its long evolution since its origin in 1884, there have been forty-five separate land transactions. New facilities were constructed in the early 1960s. A music shell was proposed, and an amphitheater for outdoor concerts was built. In 1961 the Cyr Arena, an elaborate sports complex featuring indoor skating, was erected. In 1963 a small Kiddieland

Zoo opened to allow small children close contact with animals. The older, large zoo was no longer regarded as effective.

Along with the need to accommodate the automobile, the relationship to nature could not be ignored. As a result of earlier interventions, Forest Park was beginning to face serious environmental problems such as erosion, siltation, and eutrophication in the built ponds. Vandalism was becoming another kind of problem in the park. The need for managing the complexity of the park's natural and built environment poses enduring challenges for the next era.

Environmental Education: 1966 to the Present

Confronted with a host of environmental and social problems, administrators in Forest Park began competing for more help in funding from both the state and federal governments. During the 1970s, new master plans for the park and for a new zoo were proposed. But the plans for a large regional zoo were abandoned because the threats of increased pressures from visitation and enormous parking demands seemed overwhelming.

The most interesting new role to evolve for the park since the 1960s has been its formal connection to education. Starting in 1970, the park has been utilized as an important tool in the broader educational mission of the city. An educational program for all fourth-, fifth-, and sixth-graders in Springfield began then, for the purpose of promoting environmental awareness and stewardship. It has been located in the Porter Lake boathouse, renamed the ECOS Center. The children have participated in nature study and have learned about general ecology and survival techniques in the wilds of Forest Park. A bit of Forest Park history is included in one unit. Typically, the students might go exploring ponds or collecting insects. Funds have also supported camping experiences.

While the ECOS program was starting, vandals who continued to plague the park targeted pavilions and pagodas, the very last of the rustic relics from the Pleasure Ground era. The ECOS Center program has been promoting respect generally for nature and specifically for the park; it uses education as a weapon against park vandalism. Stewardship is a key objective. It is noteworthy that the park as a natural resource is by the same token recognized as a critical educational resource.

In the mid-1980s, Forest Park was selected for inclusion in the state's Olmsted Historic Landscape Preservation Program. It could be argued that the park exhibits an Olmstedian spirit and influence, although Olmsted never had a direct hand in its design. While conducting the historic research for this program, this author could not find any previously compiled complete park history. As an historic resource within the city, Forest Park had been overlooked until 1986.

Administrators continue to operate with financial stringency. Current management practices favor the preservation of the natural stands of vegetation over any elaborate, exotic ornamental collections. Managing the natural systems is more cost-effective.

Conclusion

The idea of the park as part of an educational resource continues to offer future direction and new hope. Forest Park's past roles provide much inspiration for future uses. The education program should include earlier, more sustainable management practices of the park from the Pleasure Ground and Public Service eras. The use of native materials in constructing structures in harmony with the park's original woodland character and the preservation of native plant materials can offer important lessons today. The Public Service era especially offers compelling lessons on managing and utilizing natural resources as products. The differences between managing a natural woodland and tending an aquatic garden of exotic ornamentals could be taught. By promoting the sense of more sustainable management practices, a greater appreciation for nature could be achieved. The lesson from the Enlarged Facilities era is to caution against alienation from nature and overemphasis on the automobile.

In the Pleasure Ground era, the relationship to nature was the least complex and the most intimate. In the Social Good era, the relationship continued to be supportive. During the Public Service era, it was the most resourceful. In the Enlarged Facilities era, it was the most distant; and today, it is the most complicated. More sustainable practices throughout the system of public open spaces might be conceived to address the broader challenges that the city now faces. A rediscovery of the past solutions that helped the needy with food production and heating might be of some help in the formulation of new programs.

New roles for Forest Park in addressing such problems as homelessness, unemployment, or racism might be possible. The responses to problems in the past and the solutions found through the changing roles of an urban park can offer insight for charting the current course. From a rethinking of the roles of an urban park, a reenvisioning of urban problems is possible. Although the solutions from the past will not provide specific answers for today, history does suggest a hopeful potential.

References

Chadwick, G. E. 1966. *Park and town.* New York: Praeger.

Cranz, G. 1982. *The politics of park design.* Cambridge, MA: MIT Press.

French, J. S. 1973. *Urban green*. Dubuque, IA: Kendall/Hunt.

Marston, A. 1985. *Forest park: Bibliography and inventory*. Olmsted Historic Landscape Preservation Program, Commonwealth of Massachusetts, Department of Environmental Management. Boston.

———. 1986. *A landscape history of forest park*. Olmsted Historic Landscape Preservation Program, Commonwealth of Massachusetts, Department of Environmental Management. Boston.

Municipal reports of the City of Springfield. 1885–93. Springfield, MA: Clark W. Byran Co.

Park commissioners' reports. 1892–1930. Springfield, MA: Springfield Printing and Binding Co.

Purcell Associates. 1980. *Forest park: A revitalization plan*. Hartford, CT: Purcell Associates.

Springfield Daily Republican. 1871. Springfield and how to improve it. 6 June.

Springfield, Massachusetts, Planning Board and the Technical Advisory Corporation, Consulting Engineers, Frederick Law Olmsted, Special Advisor. 1924. *A city plan for Springfield, Mass*. Springfield, MA: Springfield Printing and Binding Co.

Springfield Sunday Republican. 1885. Forest Park. 19 July.

———. 1885. Through Forest Park. 29 November.

IV Collaborative Efforts ▬▬

The Greening of Federal Flood-Control Policies:

The Wildcat–San Pablo Creeks Case

Ann L. Riley

Introduction

Local flood-control projects along streams and rivers have long been dominated by designs that use concrete and hydraulic engineering that is rectilinear and non-ecological. However, an urban flood-control project may be designed to restore natural habitat and enhance the environment while reducing the threat from floods. The thirty-six-year history of planning for the Wildcat and San Pablo creeks in North Richmond, California, may result in an environmentally sensitive flood control project if the current design is successfully implemented and managed over time. This experience offers many lessons for similar projects elsewhere.

North Richmond is an impoverished, unincorporated community in Contra Costa County on the eastern shore of San Pablo Bay, a northern extension of San Francisco Bay (fig. 1). North Richmond grew up during World War II when blacks who came to work in the shipbuilding industry were segregated on the floodplains of the Wildcat and San Pablo creeks. Flooding puts North Richmond under a foot of water about once every three years (U.S. Army Corps of Engineers 1985). The community's need for flood control has never been disputed. However, the problems inherent in federal policies regarding the design and funding of flood control projects have repeatedly delayed construction. In the past, North Richmond residents have initiated herculean efforts to overcome federal obstacles to funding flood control projects in poor communities. They designed projects that incorporated local goals of economic recovery and environmental quality, but it required over thirty years of struggling with the federal process to get a multiobjective project constructed.

North Richmond has been considered by the U.S. Department of Housing and Urban Development (HUD) to be one of the poorest communities in the country, and therefore deserving of federal assistance. For example, the 1980 census classified 64.5 percent of the households in North Richmond as female-headed and below the poverty level. However, suburban development in other parts of Contra Costa County has made the county as a whole one of the wealthiest in California.

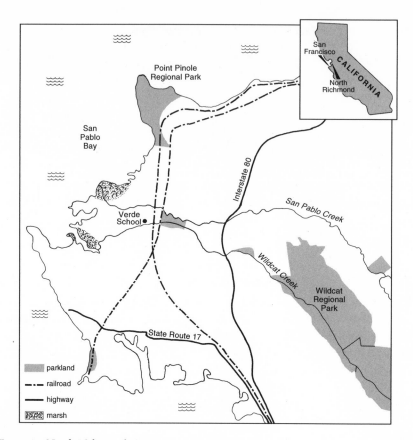

Figure 1   North Richmond vicinity

Economic redevelopment and improvement in the standard of living in North Richmond are not likely to be achieved without a flood control project. Although the community has atypical demographics because it is composed mostly of minorities, the residents' values and goals reflect those of other communities. They want opportunities, options, and environmental quality, and they want to have influence in the decisions that affect them. If North Richmond's need for flood control has been met by the federal planning process only through the greatest exertions on the part of the community, then something is wrong with federal policies and practices.

Early Efforts

In the 1940s and early 1950s, flooding along the Wildcat and San Pablo creeks attracted attention to North Richmond's need for flood control. By 1956, the Contra

Costa County Flood Control District had issued a report calling for the implementation of a flood control project. The 1960 Flood Control Act authorized the U.S. Army Corps of Engineers to conduct a feasibility study for flood control on the two creeks. At the time, the standard practice for reducing flood damage was to construct costly and environmentally damaging reservoirs and stream channels that carried more water at a higher velocity than could be carried by the natural channels.

In 1968, the Corps of Engineers issued a report that presented several different plans for flood control, but no plan was recommended for implementation because the foreseen benefits of the project did not pass the federal cost-benefit test. The only benefits the federal government recognized in its benefit-cost analysis are the value of the structures that would be protected from floods. In North Richmond, the substandard housing—some of it just cardboard boxes—was not valuable enough to justify a project.

Multiple-Objective Planning in the 1970s

In 1966 Congress approved the Model Cities Program for urban renewal. By 1971 a plan for North Richmond had been developed that identified Wildcat and San Pablo creeks and the San Pablo Bay shoreline as recreational and commercial resources that could serve as the foci for redevelopment (fig. 2). Pursuant to the Richmond Model Cities Plan, HUD commissioned a privately prepared economic analysis of a proposed flood control project in the area (INTASA 1971). Eleven years after the first federal studies began, political momentum succeeded in overcoming the difficulty of the cost-benefit analysis. HUD's consultants took into consideration future project benefits, including potential recreational benefits, and were able to justify the cost of the project.

Accepting the new favorable benefit/cost ratio, the Corps of Engineers initiated a planning process involving public participation and produced a new, community-supported flood-control plan that was authorized by Congress in 1976. A case study written on this phase of the Wildcat-San Pablo flood-control project, *Can Organizations Change?*, praised the Corps's first efforts to accommodate the needs of a poverty-stricken area (Mazmanian and Nienaber 1979). The Corps based its planning on the multiple objectives of the Richmond Model Cities Plan, which focused on social well-being, environmental quality, and economic redevelopment. The benefits of the project included the protection of existing and future development, the expected increase in the market value of the project area, and opportunities for recreation. North Richmond residents involved in planning the project during this era commended the Corps's methods and its sensitivity to community needs (Vincent et al. 1983).

The "Recommended Plan" adopted by the Corps of Engineers in 1979 contained

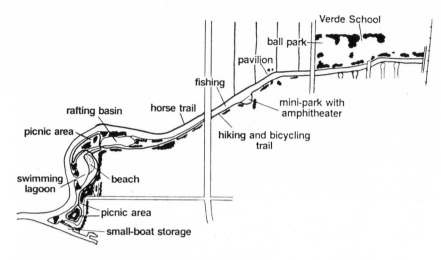

Figure 2   Richmond Model Cities Plan for Wildcat Creek. (Source: Joint Agency Committee for the Development of North Richmond-San Pablo Bay Area. "North Richmond-San Pablo Bay Area Study—Summary Report" [Contra Costa County, September 1971])

traditional flood-control engineering for the 100-year flood in the form of concrete box culverts, and trapezoidal and rectangular concrete channels. The plan also provided for an earthen, trapezoidal channel on the lower Wildcat that would have some landscaping along its sides. Also authorized as part of the project were several recreational elements, including a regional trail, a nature study area near Verde Elementary School, and freshwater impoundments on ponds (U.S. Army Corps of Engineers 1973, 1977, 1979).

Federal policy required that all land acquisitions, easements, rights-of-way, and up to 50 percent of the recreation components be paid for by the community. When North Richmond set about raising its share of the expense, some of the area's major businesses—Chevron Oil, the Southern Pacific Railroad, the Santa Fe Railroad (which had a train derail over San Pablo Creek in a January 1982 storm)— and the Richmond Sanitary Company did not contribute. Their parsimony contributed to the community's failure to raise the required local share of the total cost. Thus, the federal cost-sharing requirements undermined the Corps's efforts to design a plan that would use the creeks as part of the community economic revival plan, as outlined in the Richmond Model Cities Plan.

Under the Reagan Administration

In the 1980s, federal policies reverted to favoring the construction of projects based on the single objective of economic efficiency. The administration also required local residents to pay a greater portion of the project costs in addition to the costs of land acquisition, easements, and rights-of-way.

In 1982, Contra Costa County officials proposed a basic, structural flood-control project with no environmental amenities, to be constructed in cooperation with the Corps of Engineers. The county board of supervisors, as the local sponsor, presented the "Selected Plan" to the North Richmond community on a take-or-leave-it basis and argued that it was the only affordable option (fig. 3, a). Although open to project alternatives, the Corps decided to take a back-seat role and defer to the county on the issues of project design and citizen participation. The Corps of Engineers also discouraged multiobjective planning on the assumption that North Richmond could not afford anything but a basic channelization project.

Some North Richmond residents were resigned to accepting any flood control project offered. Others felt so strongly about the Richmond Model Cities Plan that they wanted to retain influence in the design process and explore other project options. The take-it-or-leave-it offer ran contrary to the long history of active community involvement in the Richmond Model Cities Plan and alienated some key community leaders. In the spring of 1983, these leaders organized a meeting in North Richmond to determine community reaction to the county/Corps Selected Plan for flood control. The issues raised at that meeting defined the next five years of work for the community volunteers, who changed the planning process, the plan design, and funding strategy.

Several North Richmond community groups—the Urban Creeks Council, the Save San Francisco Bay Association, and the Contra Costa County Shoreline Parks Committee—formed a coalition. They collectively developed their own plan that recognized the value of Wildcat and San Pablo Creeks as important local and regional resources. This plan raised several important environmental concerns:

Wildcat Creek was classified by the California Department of Fish and Game as one of the last remaining streams in the San Francisco Bay area with an almost continuous riparian environment along its length. However, the county/Corps Selected Plan would make the creek into a concrete and earth-lined channel, complete with covered box culverts.

Environmental experts, including two nationally prominent hydrologists, Luna Leopold and Philip Williams, feared that sedimentation caused by the project would seriously harm the wetlands and marshes of the lower floodplain. Hydrologists reported to the coalition that the Corps's estimates of the amounts of sediment moving through the two creeks were substantially low. The concrete-lined channels would therefore not provide the flood protection assumed by the project's designers because the sediment would increase the hydraulic resistance and decrease the capacity of the channels. The plan would thus require costly and fre-

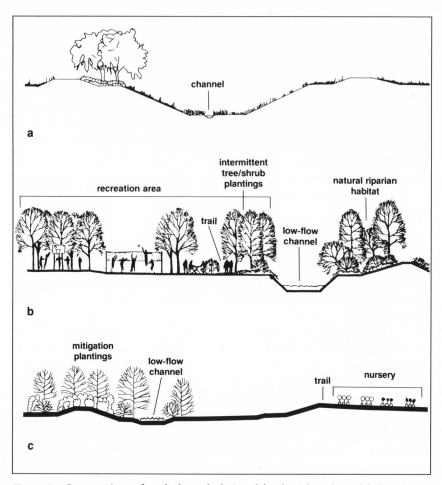

Figure 3 Cross sections of creek channels designed for the Selected, Modified, and Consensus plans for the flood-control project on Wildcat and San Pablo creeks in North Richmond, California (Sources: U.S. Army Corps of Engineers, *General Design Memorandum and Basis of Design for Reach 1, Wildcat and San Pablo Creeks* [Sacramento, Calif.: U.S. ACE, Sacramento District, October 1985]); Poster of the Modified Plan published by a coalition of North Richmond community organizations including the East Bay Regional Park District; and U.S. Army Corps of Engineers, *Supplement No. 3 to Design Memorandum 1, Wildcat and San Pablo Creeks Environmental Mitigation Project* [Sacramento, Calif.: U.S. ACE, Sacramento District, August 1988]): (a) 1982 Selected Plan proposed by Contra Costa County and U.S. Army Corps of Engineers; (b) 1984 Modified Plan proposed by a coalition of North Richmond community organizations; (c) 1986 Consensus Plan developed by a design team of community, county, and federal representatives.

quent maintenance, and the proposed sediment detention basin on Wildcat Creek would not protect the marshland of the lower floodplain from sedimentation. There were no sponsors or plans to provide recreational open space and educational benefits for members of the community and regional park users.

The Urban Creeks Council and the Richmond Neighborhoods Coordinating Council decided to design their own flood-control plan and successfully applied to the Vanguard Foundation in San Francisco and the San Francisco Foundation for funding. The coalition of neighborhood and environmental organizations, using a 1960s strategy for community participation known as advocacy planning, solicited its own experts to develop a new plan to compete with the county/Corps Selected Plan.

The Modified Plan

The East Bay Regional Park District had long sought the extension of popular regional trails from Wildcat Canyon and Point Pinole Shoreline parks along Wildcat and San Pablo creeks and their marshes. Financial assistance from the park district and the Save San Francisco Bay Association brought the coalition's final budget for alternative planning to $50,000, enough to pay for the design of a flood control project on at least one of the creeks, while the principles and many of the details of the design would be applicable to both creeks. Eventually, the citizens developed a "Modified Plan" for Wildcat Creek that had a very different design philosophy from that of the county/Corps Selected Plan (Williams and Vandivere 1985). This new plan would modify the existing creek channels to simulate the natural hydraulic shape and processes of undisturbed streams; the altered channels would deposit the sediment in the upstream floodplain, and help to restore valuable riparian vegetation. The proposed concrete and trapezoidal earth channels of the Selected Plan would be replaced with more natural, low-flow, meandering channels lined by floodplains, setback levees, planted gabion walls, and riparian trees (fig. 3, b). The Modified Plan also included regional trails and educational and park facilities. The coalition's planners developed their own project cost estimates and funding strategy and presented their Modified Plan at the same meetings where the Selected Plan was presented to the public and government agencies.

The advocacy planning strategy introduced alternatives, and therefore controversy, into the Corps of Engineers' planning sessions. The strategy eventually forced a change in the planning process, from one in which citizens were to be briefed on the final plan chosen by the county board of supervisors and the Corps of Engineers, to one in which citizens became active participants in determining the design of the final plan.

On 19 February 1985, the Contra Costa County Board of Supervisors approved the Selected Plan for construction, but left the door open for multiobjective designs if funds became available. In June 1985, the U.S. Fish and Wildlife Service (USFWS) reviewed the proposed plans in terms of their probable impacts on the marshes and their endangered species. It selected the coalition's Modified Plan as "the prudent and reasonable alternative" (U.S. Fish and Wildlife Service 1985). In addition, the San Francisco Bay Conservation and Development Commission (BCDC) did not find the Selected Plan consistent with the requirements of the McAteer-Petris Act for the protection of the San Francisco Bay wetlands. (The act requires the regulation of fill placed in the bay and requires public access to the bay.) But BCDC did approve the Modified Plan. A combination of pressure from federal and state environmental and regulating agencies, the endurance and persistence of community leaders, and press coverage finally resulted in the adoption by the Contra Costa County Board of Supervisors of a multiobjective "Consensus Plan." Construction of the Consensus Plan began in 1987.

Design by Consensus

When the Corps of Engineers found in June 1985 that it could not implement the Selected Plan, the county board of supervisors established a project design team to construct a plan in which the concerns of the government agencies with regulatory powers over the project would be properly coordinated and integrated with the concerns of the public. The team, which met at least monthly, included representatives of the Corps of Engineers, USFWS, and relevant state and regional agencies, including the California Coastal Commission and BCDC. Local political officials, landowners, and commercial interests were also involved. The team was to produce a fundable project that the regulatory agencies would accept and the coalition could endorse.

Two notable problems arose. First, some relevant and interested parties were not included on the design team. Second, continuity in decision making and plan formulation broke down because of frequent changes in Corps and county staffing.

The first problem occurred because the Richmond Unified School District Board was not adequately involved in the design project, which ran through their property near Verde Elementary School. The school board held up the project by withholding the right-of-way until its concerns were met. By withholding the right-of-way, the school board was able to force a more environmentally sensitive treatment of the part of the creek running through school property.

The second problem that plagued the design team was the lack of continuity in both federal and local staffs assigned to the project. Between 1984 and 1988 the Corps of Engineers assigned three different engineers to the job of project manager. The resultant discontinuity in decision making brought on an environmental and publicity disaster featured on the front page of the *San Francisco Examiner-Chronicle* on 14 June 1987 (Brazil 1987). Construction plans that did not reflect the decisions of the design team were given to the contractors, who accordingly bulldozed a half-mile of riparian vegetation that was supposed to be preserved. Shortly thereafter, a levee constructed in the wrong location prevented the implementation of a marsh restoration project and jeopardized state funds for the marsh enhancement plan. The situation was exacerbated when a key member of the county staff gave the contractors approval to proceed with plans that did not correspond to the team's decisions.

Control Features

The design team chose features from both the Modified and Selected plans. Although the team's final Consensus Plan is a compromise between the two earlier plans, the basic components of the Modified Plan were retained, in recognition of the importance of managing the large amount of sediment, particularly in the Wildcat watershed, to avoid degrading endangered species' habitat in the marshes (fig. 3, c).

One of the most important features of the coalition's Modified Plan was that the stream corridors, or floodways, would remain within the same narrow right-of-way boundaries proposed by 1982 county Selected Plan and would offer the same level of protection against a 100-year flood. The rights-of-way in the Corps of Engineers' original 1976 plan had been up to 250 feet wide to accommodate certain environmental features. The Modified Plan included riparian vegetation next to the channels and a terrace for sediment accumulation but did not increase the project's width beyond 180 feet. Yet the designs of the Modified Plan that were incorporated into the Consensus Plan provided the same level of flood protection as the 1976 design. A different design philosophy was used in which the channels were modeled not on the dimensions of a hydraulic flume, but rather on the geometry of natural channels. Thus, the design of the Consensus Plan disproves the common assumption that only trapezoidal or rectangular channel geometry can be used in a narrow right-of-way.

The Consensus Plan replaced the standard trapezoidal dirt and riprap channels, rectangular concrete channels, and box culverts of the Selected Plan with natural floodplain features of the Modified Plan wherever possible. The Consensus Plan has ten- to fifteen-foot wide, meandering, low-flow channels designed to carry the creek's average flow, and floodplains where higher flows could spread, lose velocity, and deposit sediment. Riparian vegetation is included on both sides of the low-flow channels. Riparian trees will shade the channels and prevent the growth of bulrushes and willow, which obstruct flow. Although previous Corps project designs had designated a low-flow channel in lower Wildcat Creek, they had not included natural channel geometry, vegetation, or grading plans that would help define stable, low-flow channels. Typically, the Corps's low-flow channels, superimposed on open, wide-bottomed, trapezoidal corridors, are unstable, braided, and choked with bulrushes.

The Consensus Plan provides that sediment will be deposited where it is least harmful—on the floodplain and in the bay. By trapping sediment in the upstream floodplains, filling of the downstream marsh with sediment should be prevented. The Consensus Plan assures that the low-flow channels will scour as much sediment as possible and transport it to San Pablo Bay. To further protect the marsh from sedimentation, the plan also calls for widening the slough channels through the marsh so that suspended sediments can be conveyed without overtopping into the marsh, and for excavating sediment to increase the brackish area and restore the marsh's tidal action.

Technical Issues

The most contentious technical issues faced by the design team included making reasonable estimates of the sediment loads carried by the creeks, assessing the ability of the Corps's proposed sediment basin to collect sediment, judging the safety of concrete box culverts, and assigning Manning's Equation (roughness values) to areas proposed for revegetation. The coalition's experts argued that the natural creek channels were aggrading with high sediment loads and predicted that the even wider, trapezoidal channels proposed by the Corps would further increase sedimentation. The narrow, low-flow channels of the Modified Plan were better designed to transport sediment in suspension at higher velocities. The design team decided to locate the basin farther upstream, and adopted the floodplains, the wetland transition zone, and the higher velocity, low-flow channels of the Modified Plan to keep the sediment load from ending up in the marsh or significantly decreasing the channel's capacity.

Another difficult design issue to resolve was how to make up for the loss of twenty-four acres of riparian vegetation. The county's 1982 proposal called for planting trees on some land north of Wildcat Creek. In the Consensus Plan, trees planted along the low-flow channels of the two creeks would help guide channel formation and shade the banks to prevent them from clogging with rushes, reeds, and sediment. Thus, choosing roughness values that would determine how much vegetation could be allowed without reducing the needed channel capacity became a critical aspect in the design of the Consensus Plan.

Once roughness values had been chosen, the design team had to agree upon a maintenance plan for keeping the low-flow channels cleared of vegetation until a riparian canopy could grow to shade out the unwanted, clogging reed growth. The agreement negotiated between the county supervisor and the Corps project manager provides for inexpensive hand labor by conservation crews to clear the unwanted vegetation. Maintenance crews have included the State of California Conservation Corps and a local East Bay Conservation Corps as well as labor from the state's new work force program. It was also agreed that the standard, annual maintenance routines for removing sediment or clearing of vegetation would be replaced by a maintenance schedule based on actual need. Thus, maintenance costs and negative environmental impacts resulting from channel maintenance should be reduced.

In order to design a suitable revegetation plan, the county asked the Corps of Engineers to contract with the Soil Conservation Service (scs), which has experience with the revegetation and restoration of streams. In September 1988, the scs and the Corps of Engineers issued a recreation and revegetation supplement to the Corps's design memorandum about the Consensus Plan (U.S. Army Corps of Engineers 1985). They proposed not to landscape a flood control project but to restore the riparian environment along the low-flow channels. Revegetation will be done with cuttings from nearby plants, seeds from California species native to the locale, and some container stock. The competence demonstrated by the landscape architects in the design process inspired the design team to ask the Corps to retain the scs staff for the actual plant installation.

Funding Strategy

The coalition's Modified Plan and the county's Selected Plan had very similar cost estimates. The Consensus Plan's costs were higher because the sediment basin was to be redesigned and relocated. The transition of this project from a single-objective flood control project to a multiobjective project to restore marshes, provide recreational and educational opportunities, and enhance the environment, as well as to control flood damage, made it possible to attract funding from state agencies that could not otherwise have contributed. For example:

—The East Bay Regional Park District committed $793,000, matched by another $793,000 from the Corps of Engineers, for a regional trail system. The park district later committed $19,000 to help enhance creekside educational opportunities near Verde Elementary School and may commit more as the recreational and educational elements are finalized.

—The California State Lands Commission spent $240,000 on land for the Wildcat Creek wetland transition zone.

—In February 1987, the California Coastal Conservancy Board authorized $578,000 for marsh restoration and riparian enhancement areas. After the original restoration plan was damaged by the construction mistakes in the Wildcat and San Pablo creek marshes and the county failed to identify willing sellers of riparian land parcels, a task force headed by the Coastal Conservancy prepared a new marsh restoration plan.

—In June 1989, the California Department of Water Resources awarded a $100,000 grant because the project involved design innovations, a commitment to citizen participation, and educational opportunities.

—In 1992 the Near Coastal Waters Program of the Environmental Protection Agency, the San Francisco Foundation, and the San Francisco Estuary Program contributed approximately $35,000 to revegetation and education projects.

Policies and Practices

Federal policies for project evaluation and funding are strongly biased against a project like North Richmond's. Federal definitions of water project costs and benefits do not reflect the broad, long-term needs and values of the communities where such projects are often located. Likewise, federal cost-sharing policies do not recognize local socioeconomic conditions. Federal policies discriminate against financially disadvantaged communities attempting to benefit from federal flood-control projects, even though these communities are frequently located in some of the most hazardous areas. Because the cost-sharing policies make it a local responsibility to purchase lands, easements, and rights-of-way, they discourage the purchase of riparian preservation zones, trails, and other environmental amenities.

In the interest of holding down federal water projects expenditures, federal cost-benefit analysis and cost-sharing requirements are biased against low-income areas and nonstructural solutions. Even the environmental lobby supports the federal cost-sharing policies, in the belief that such policies will reduce the number of projects and thus reduce damage to the environment. It is contradictory for environmental advocates to challenge the use of the cost-benefit analysis as an oversimplified means to justify the selection of projects for federal assistance, but to support the use of cost-sharing arrangements that may defeat desirable projects in disadvantaged communities.

Cost-benefit analysis and cost sharing should not be the only determinants in qualifying projects for federal support. Local priorities and objectives must be incorporated into the plans, as should broader national goals for meeting social and environmental needs.

A reformed system using objectives-based planning and technical designs based on concepts of hydrology instead of channel hydraulics would reduce both the federal share of costs and the total bill for project construction. Objectives-based planning will save federal dollars because (1) the projects that will legitimately meet the test of fulfilling multiple objectives are few; (2) alternative strategies such as stream restoration can lower project costs; (3) alternative construction and maintenance techniques may contribute to local economies just as the Works Progress Administration did in the 1930s and 1940s; and (4) protection against the smaller, more frequent floods instead of the larger, 100-year floods will reduce the costs of many projects.

Citizen participation is considered by many water project planners to be a costly nuisance, but many project engineers and members of Congress can tell of dramatic overruns in planning costs that occurred following years of studies and planning when citizens blocked projects after they were authorized or before construction started. Most federal water-project planners do not realize that a high level of citizen participation can attract financial contributors to projects. Citizen participation can also stimulate political support and interest in a project, and in turn attract money from a number of local, county, regional, and state programs. In addition, just as the multiple objectives of the Consensus Plan brought in nonfederal funds, projects that have objectives in addition to flood control, such as park development, fisheries enhancement, recreation, and wildlife benefits, save federal dollars by attracting other funding sources, such as state and local resource, fish and game, and park agencies.

Some nonstructural and environmentally sensitive design measures do incur higher costs for land acquisition. But these costs need to be balanced against the long-term costs of maintaining structural engineering works, constant sediment removal, vegetation removal, and the unintended impacts common to the traditionally designed projects.

The U.S. Army Corps of Engineers is proud of the flood control project on Wildcat and San Pablo creeks. An engineer for the Sacramento Water District wrote an article for *Hydraulic Engineering* describing the interesting hydraulics of the Consensus Plan (Sing 1988). The Corps's Waterways Experiment Station (WES) has encouraged the use of this project as a model for future water-project designs in training courses. Recently, WES adopted the Wildcat Creek Project as logo for its national Flood Control Research Program. However, well-intentioned Corps personnel who want to respond to local needs in formulating plans find themselves caught in a conflict between local needs and federal policies. The project in North

Richmond is just one of twelve California water projects that the public has tried to redesign to meet community needs over the last ten years.

The current federal system of water project evaluation is so narrow that only those communities with the most influential representatives will be able to undertake the long and costly process necessary to circumvent the system and get a project that meets community needs. Federal water policies have a tendency to provide well-off communities with poorly designed projects and poor communities with nothing.

Acknowledgment

This article is adapted from an article with the same title that appeared in *Environment* 31, no. 10 (December 1989):12–20, 29–31. It is reprinted with permission of the Helen Dwight Reid Educational Foundation. Published by Heldref Publications, 1319 Eighteenth St. N.W., Washington, D.C. 20036-1802. Copyright © 1989.

References

Brazil, E. 1987. Trouble anew for "Murdered" Creek. *San Francisco Examiner-Chronicle*, 14 June.

INTASA. 1971. *Relationship of proposed flood control project and model cities objectives for community development in North Richmond*. Menlo Park, CA.

Mazmanian, D. A., and J. Nienaber. 1979. *Can organizations change?* Washington: The Brookings Institution.

Sing, E. F. 1980. Stable and environmental channel design. *Hydraulic Engineering*. August.

U.S. Army Corps of Engineers (USACE). 1973. *Wildcat-San Pablo creeks, Contra Costa County, California, feasibility report for water resources development*. San Francisco: USACE.

———. 1977. *Wildcat-San Pablo creeks, Contra Costa County, California, general design memorandum phase 1 for flood control and allied purposes*. Draft. San Francisco: USACE.

———. 1979. *Master plan Wildcat-San Pablo creeks*. Plan prepared by Arbegast, Newton & Griffith, landscape architects, Contra Costa County, CA. Draft. San Francisco: USACE.

———. 1985. *General design memorandum and basis of design for reach 1, Wildcat and San Pablo creeks*. Sacramento: USACE.

U.S. Fish and Wildlife Service. 1985. *Endangered species formal consultation on the proposed San Pablo and Wildcat creek flood control project, Contra Costa County, CA*. AFA-SE 1-1-85-F-19. Portland: U.S. Fish and Wildlife Service.

Vincent, B., J. Vincent, L. Edwards, L. Hunter, J. Siri, and members of the U.S. Army Corps of Engineers planning committees. 1983. Conversations with the author.

Williams, P., and W. Vandivere. 1985. *A flood control design plan for Wildcat and San Pablo creeks*. Sponsored by the San Francisco Foundation, the Vanguard Foundation, the East Bay Regional Parks District, and the Save San Francisco Bay Association. San Francisco.

Reconciling Urban Growth and Endangered Species:

The Coachella Valley Habitat Conservation Plan

Timothy Beatley

The Endangered Species Act

The Endangered Species Act (ESA) (16 U.S. Code, secs. 1531–1543) represents the cornerstone of federal efforts to protect endangered flora and fauna. Originally enacted by Congress in 1973 and reauthorized most recently in the fall of 1988, the act sets forth a strong national mandate to protect and manage endangered species. Under the provisions of the act, flora or fauna can be listed either as endangered or threatened, and where appropriate, critical habitat is designated. There are more than seven hundred U.S. plant and animal species listed as endangered or threatened and several thousand additional species listed as candidates. Once they are placed on the list, the ESA provides special protection to these species and their habitat (see Defenders of Wildlife 1987; Yaffee 1986).

A primary protective element of the ESA is found in Section 9, which prohibits the "taking" of any listed species. The term "take" is defined broadly in the act as including "to harass, harm, pursue, hunt, shoot, wound, kill, trap, capture, or collect, or to attempt to engage in any such conduct." (Endangered Species Act, Section 3[19]). Section 9 has been particularly problematic for land developers who could be subject to criminal and civil penalties for undertaking grading, land clearance, and other construction-related activities that might harm or kill listed species. However, it has always been difficult for the U.S. Fish and Wildlife Service (USFWS) to enforce the Section 9 provisions, which virtually require agents to find actual dead animals before prosecuting developers.

Habitat conservation plans under the Endangered Species Act have been prepared or are in the process of being prepared in a number of locations around the country. The second habitat conservation plan to be formally approved by the U.S. Fish and Wildlife Service was that drawn up for the threatened Coachella Valley fringe-toed lizard. This plan has been profiled in such publications as *National Geographic* and the *New York Times* and on television shows such as the "ABC Nightly News," and is widely heralded as a national model for resolving development/conservation conflicts. What follows is a detailed case study of the Coachella

Valley Habitat Conservation Plan (HCP) that discusses its history, key provisions, success to date, and implications for public policy. The study begins by reviewing the types and magnitude of the development pressures in the Coachella Valley that have given rise to this conflict.

The Coachella Valley

The Coachella Valley lies approximately one hundred miles east of Los Angeles and is a northern extension of the Colorado Desert. The valley is three hundred square miles in size and is bounded to the north by the Little San Bernardino Mountains and to the south and west by the Santa Rosa and San Jacino mountains. San Gorgonio Pass is at the western end of the valley, and the Salton Sea lies at the extreme east. The valley consists largely of windblown sand and rocky alluvial deposits.

The southern portion of the valley was converted to agricultural uses in the 1940s, largely as a result of the construction of the Coachella Canal in 1948. This canal brought water to the valley and permitted for the first time large-scale agricultural activities, occurring especially in the southern end of the valley. Urban development in the postwar period was initially modest taking place mainly south of the Whitewater River. As the demand for land and housing has increased over the years, urban development has expanded into blowsand habitat areas. The expansion of roads and highways and the invention of air conditioning have made the desert environment more accessible and attractive.

The valley has experienced tremendous growth in recent years. While the permanent population was only about 12,000 in 1940, it rose to more than 130,000 in 1980, and is projected to increase to 311,000 by the year 2000. (Coachella Valley Association of Governments 1988, 1989). Because much of the population in the valley is seasonal, population pressures are even greater than these estimates suggest. Several of the fastest-growing cities in California are located in the valley, including Desert Hot Springs, Palm Desert, and Rancho Mirage (table 1). Much of the valley has developed as an exclusive resort area, and is home to numerous entertainers and celebrities, including such notables as former President Gerald Ford, Frank Sinatra, Bob Hope, and others. As development pressures have increased over the years, much of the natural desert environment has disappeared. This has in turn meant shrinking habitat for wildlife, both endangered and nonendangered, and an ever greater intrusion of man into the natural landscape.

The Endangered Lizard

The Coachella Valley fringe-toed lizard (*Uma inornata*) is a medium-sized lizard, averaging six to nine inches in length (Coachella Valley Steering Committee, Sec-

Table 1   Population Changes in the Coachella Valley (permanent population)

	1970	1980	1988	Projected 2000
Palm Springs	21,497	32,366	40,925	65,601
Cathedral City	7,327	11,096	26,758	29,847
Desert Hot Springs	2,738	5,941	10,383	17,150
Rancho Mirage	2,767	6,281	8,525	16,077
Palm Desert	6,171	11,801	18,088	30,053
Indian Wells	760	1,394	2,443	6,348
Indio	14,459	21,611	33,068	47,945
La Quinta	1,190	3,328	9,274	17,908
Coachella	8,353	9,129	14,115	23,625
Coachella Valley (including unincorporated areas)	88,999	133,419	202,231	311,911

Source: Coachella Valley Association of Governments (1988, 1989).

tion II). Whitish or sand-colored overall, it exhibits a pattern of eyelike markings forming longitudinal stripes along its shoulders. (Cornett 1983; Carpenter 1963.) The lizard has morphological features and behavioral patterns that are distinctly suited to life in the desert. In many ways, the lizard is the quintessential example of successful evolutionary adaptation. Its most obvious features are the fringes on its toes—a row of scales on the bottom edge of the toes that substantially increases the foot surface and allows the lizard to "skate" along the sand, often at high speed. The lizard also can dive into sand and move under sand for short distances—what is sometimes called sand swimming. Other adaptations include smooth scales to reduce friction, an ability to partially close its nostrils to keep out sand, a U-shaped nasal passage that serves to trap sand and allows the lizard to blow out these trapped materials easily, and a wedge-shaped snout to facilitate sand diving. The lizard also has fringed eyelids with a double seal, and a special flap of skin that covers its ears when it dives into sand. The upper jaw overlaps the lower jaw, which helps to prevent the entrance of sand when it dives underground. Its behavior is also specially adapted to life in its desert environment (Cornett 1983).

The Coachella Valley fringe-toed lizard lives in areas of fine windblown sand and thus relies heavily upon a natural, uninterrupted "blowsand ecosystem." (A blowsand ecosystem is one in which important habitat is created through the deposition of windblown sand. These blowsand habitat areas include dunes, drifts, and sandy plains.) Sand materials originate in the mountain areas and are brought into the valley through periodic flooding. Strong winds pass through San Gorgonio Pass at the northwestern end of the valley and move in a southeastern direction, transporting materials and depositing them in alluvial fans down the valley. These strong winds from the northwest also serve to sort the materials.

At one time the range of the Coachella Valley fringe-toed lizard extended throughout the valley. Its range has been slowly reduced since the turn of the century as a result of a combination of factors, including the conversion of land to agricultural uses, the interruption of natural sand movement by buildings, railroad windbreaks, and other human-made improvements, and the loss of habitat to urban development (i.e., roads, houses, golf courses). Off-road vehicle use has also taken its toll on the lizards.

Of the original 267 square miles of habitat, it has been estimated that one-half has been directly converted to other uses. As of the completion of the habitat conservation plan in 1985, the occupiable habitat of the lizard was estimated to be approximately 127 square miles, or 81,500 acres. Furthermore, much of this remaining occupiable habitat is currently undergoing degradation due to the disruption of natural sand transport needed to sustain the blowsand habitat. It has been estimated that as much as 50 percent of the remaining occupiable habitat is currently experiencing some degradation due to blowsand shielding. In addition to roads, buildings, and other forms of urban development, various measures to control blowing sand have also resulted in shielding.

Serious concerns about the lizard's plight emerged in the mid-1970s, as studies began to show substantial declines in lizard habitat and population (see England and Nelson 1976). A group of concerned biologists and resource managers formed the Coachella Valley Fringe-toed Lizard Advisory Committee and pushed to have the lizard placed on the federal and state endangered species lists. The group also began to advocate the establishment of a desert habitat preserve to protect the lizard and other desert biota. The USFWS proposed in 1978 to list the lizard as threatened and to designate a relatively large critical habitat (170 square miles). Though political opposition to the listing was fierce, the lizard was actually placed on the federal list in 1980, but a substantially smaller critical habitat was designated (approximately 20 square miles). The State of California also placed the lizard on its own list of rare and endangered species in the same year (listed as endangered).

The Lizard versus Development

It was not until 1983 that serious conflicts emerged between urban growth and development in the valley and the protection of the lizard. Several development projects were proposed in areas of potential lizard habitat, including a 433-acre, 1,300-unit country club/condominium project near Palm Desert (actually located in Riverside County). Controversy over this project in particular directly lead to the initiation of the Coachella Valley fringe-toed lizard habitat conservation plan. While the developer denied that lizards were present on the site, local environmentalists questioned the project's biological assessment, arguing that lizard take would be likely to occur. Local environmentalists demanded that the developers mitigate the habitat loss by protecting (i.e., acquiring and setting aside) an equal number of habitat acres elsewhere in the valley. The developers objected to the high cost of such a requirement and threatened to sue the county if not allowed to proceed. An impasse quickly was reached.

The "Lizard Club"

A compromise was eventually arrived at in which both sides would participate in the development of a habitat conservation plan, which would eventually result in the establishment of a lizard preserve (Beatley 1990). A working group was formed (which became affectionately known as the "lizard club") consisting of representatives of major stakeholder groups in the region, including the Coachella Valley Association of Governments (CVAG), the Bureau of Land Management (BLM), the California Department of Fish and Game (CDFG), the U.S. Fish and Wildlife Service (USFWS), the Coachella Valley Water District, the Riverside County Planning Department, the Coachella Valley Ecological Reserve Foundation, The California Nature Conservancy, the Sunrise Development Company, and the Agua Caliente Indian tribe. A smaller steering committee was formed to develop the specifics of the habitat conservation plan, and the California Nature Conservancy took on the role of organizing and coordinating the process. A San Francisco consulting firm was hired to collect the necessary background information on biology and land use, and to prepare the plan itself and the accompanying environmental documentation.

Beginning in 1983, the lizard club met monthly over a two-year period. As expected, diverse values and concerns emerged. The key management strategy pursued by the group was the establishment of one or more undisturbed lizard preserves. Early on, the concept of on-site mitigation was discarded, as the lizard's viability was so closely tied to the existence of well-functioning natural blowsand habitat. It was generally concluded that attempts to preserve small, scattered fragments of habitat would not be successful.

Much of the group's discussion centered around how many preserves were necessary, how large these preserves would have to be to ensure the long-term survival of the lizard, and where they should be located. The group was strongly influenced by the recommendations of the USFWS Fringe-toed Lizard Recovery Plan. Under the Endangered Species Act, the U.S. Fish and Wildlife Service must prepare a recovery plan for each listed species. Among other things, these plans typically evaluate the status of a species and the threats to its existence, and establish goals and actions needed to ensure its survival and recovery. These plans are usually authored by a team of biologists with expertise on the species of concern. (For a discussion of recovery plans, see General Accounting Office 1988). Consensus formed around the recommendations of the USFWS Fringe-toed Lizard Recovery Team that at least two preserves be established, each a minimum of one thousand acres in size (USFWS 1984). Because of the importance of the blowsand ecosystem, each preserve was also to include its own sand source. Considerable energy was spent discussing the specific locations and configurations of the preserves. Eventually the group decided on the establishment of a system of three preserves—one large preserve (the Coachella Valley Preserve), and two smaller satellite preserves, each containing its own sand source.

Sharing the Costs

It was generally agreed that the costs of the preserves should be shared by both the public and private sectors. The developers argued that because the Endangered Species Act was a federal statute, it was only fair that the larger public bear a substantial portion of the costs of the preserves. Two sources of federal participation were pursued: land trades through the Bureau of Land Management, and acquisition moneys from the federal Land and Water Conservation Fund. The use of BLM land exchanges represented an ingenious approach to securing preserve lands. Under the arrangement, The California Nature Conservancy (TNC) would purchase land within the Coachella Valley Preserve, then swap these lands for BLM lands in other locations. Once TNC was given title to BLM lands outside of the preserve, it would then sell these lands on the open market, using the proceeds to repay itself for the costs incurred in buying the preserve land.

The local development community also agreed to a significant contribution in the form of a per-acre mitigation fee for development taking place in habitat areas. A mitigation fee of six hundred dollars per acre was established after considerable dissension between representatives of the development industry and members of the lizard group, including environmental advocates and county planning staff (Beatley 1990). Much of the lizard club's work involved the technical and political process of establishing the boundaries of the zone where the mitigation fee would apply. It was generally decided that the boundaries of the fee zone should correspond to the historic range of the lizard. It was also decided that the Whitewater River should comprise the western boundary of the fee zone, as the lizard's habitat for the most part did not extend south or west of the river. However, the boundary did dip below the river in several areas where it was clear that potential habitat existed. The group received substantial pressures to modify the fee boundaries, usually in the interest of excluding certain lands. The lizard group responded fairly firmly to such attempts to play with the boundaries. A major argument made to the development community, with considerable effectiveness, was that the smaller the fee zone, the higher the per-acre mitigation fee would have to be. Maintaining a fairly large fee zone meant that the costs would be distributed over a greater number of landowners and development projects. Protecting the integrity of the boundary was also seen as important to ensuring the legal and scientific soundness of the plan, and preventing arbitrary differences in the treatment of similarly situated landowners. Nevertheless, some concessions were made to local governments, and the mitigation fee boundaries were modified in a few cases in response to local pressures.

For the plan to work, ten local jurisdictions in Coachella Valley would also have to sign off on the habitat conservation plan. Throughout the HCP process, periodic presentations to local city councils were made by members of the lizard club steering committee. Generally, the local governments of Coachella Valley could be characterized as progrowth and prodevelopment, and the attitude toward the lizard issue was by and large one of either indifference or antagonism. Little or no concern has been voiced about preserving the species itself—rather, the species was seen as a problem to be overcome. Considerable outright hostility was expressed toward the lizard issue by some local governments, particularly those of Palm Springs and Rancho Mirage. In the beginning of the HCP process, the city council of Rancho Mirage voted to sue the U.S. Fish and Wildlife Service over the issue. (*The Press-Enterprise* 1982a.) The mayor of Palm Springs at the time was one of the people most incensed over the lizard issue (Moore 1983): "It is the biggest racket ever perpetrated on this Valley. . . . I say racket because the biologists in this area want to get a piece of our land for all the little animals and take it away from people who have been paying taxes for years and years." The then-mayor of Rancho Mirage expressed similar feelings, calling it "ridiculous" to "stop progress because of a lizard." (*The Press-Enterprise* 1983b.) The USFWS was generally seen as a common enemy of the valley.

The need to stick together was a plea frequently made by the staff of the Coachella Valley Association of Governments (CVAG), which directly represented the local governments in the process. If any one of the localities chose to balk at the plan, it was argued that this would jeopardize the entire solution. Despite the threats by several local governments along the way to abandon ship, they all eventually embraced the plan and signed the necessary implementing agreement. For most local elected officials, this was not an issue of high importance to them or their constituents. There was certainly no groundswell of popular demand to save the lizard. To many of these officials, the whole idea of expending a large amount of money to protect such a creature was silly and preposterous (and the butt of many jokes).

Nevertheless, the attitudes of numerous local officials did change markedly over time—not, however, as a result of any greater appreciation of the lizard. For local elected officials, the more convincing arguments appeared to be those that stressed the importance of preserving the land, not the lizard. The biologists and other advocates of the lizard frequently pointed out that the preserve lands, particularly the Coachella Valley Preserve, would be likely to be viewed by residents in the future in the same affectionate way that New York City residents view Central Park. As the valley continued to develop, these lands might be the last remnants of natural open space. Elected officials eventually came to view the preserves as something important to set aside for future generations.

The final draft of the habitat conservation plan was reviewed and approved by the ten local governments in the late fall of 1984 and early spring of 1985. Once approved by each of the local governments, the plan was sent to the U.S. Fish and Wildlife Service, which approved the plan in April 1986, issuing a thirty-year Section 10(a) permit. Thus, roughly three years after the initiation of the process, a long-term conservation plan had been adopted. What follows is a detailed summary of the HCP as ultimately approved.

The Resulting Solution: The Coachella Valley Fringe-Toed Lizard Habitat Conservation Plan

Table 2 lists the total projected acreage and lizard habitat acreage for the three preserves. The largest of the three preserves—the Coachella Valley Preserve—is eventually to contain more than 13,000 acres. Located north of Interstate 10, in the Thousand Palms canyon area, this preserve will comprise approximately 5,200 acres of fringe-toed lizard habitat (fig. 1).

Two smaller "satellite" preserves, the Willow Hole-Edom Hill Preserve and the Whitewater Floodplain Preserve, were also included in the plan, each with its own sand source. The lands in these two preserves were already partially under BLM ownership or control. The Whitewater Floodplain Preserve contains 1,230 acres, all of which is lizard habitat (table 2). While the land is technically owned by the Coachella Valley Water District, it is under management control by the BLM. The Willow Hole-Edom Hill Preserve contains 2,469 acres, of which 1,407 are lizard habitat (table 2). While a large portion of this land was already owned by the BLM, the habitat conservation plan envisions the acquisition of an additional 775 acres of adjacent private land. Other BLM lands on which wind-energy leases have been issued will also be managed to protect lizard habitat. These BLM preserve lands are

Table 2   Amount of Lizard Habitat Protected in Preserves (projected acquisition)

Preserves[a]	Total Preserve Acreage	Amount of Habitat[b]	Total Lizard Habitat in Coachella (%)
Coachella Valley Preserve	13,030	5,201	6.4%
Willow Hole/Edom Hill Preserve	2,469	1,407	1.7%
Whitewater Floodplain Preserve	1,230	1,230	1.5%
Total	16,729	7,838	9.6%

Source: Coachella Valley HCP Steering Committee (1985).

[a] Estimated total lizard habitat in the valley in 1985 was 81,500 acres.

[b] Potentially occupiable habitat; this acreage is not necessarily currently occupied by fringe-toed lizards.

to be protected as areas of critical environmental concern (ACECs), a designation that prevents such activities as the use of off-road vehicles.

Together these three preserves will encompass nearly 17,000 acres of protected land, including some 7,838 acres of occupiable habitat. It is estimated that this represents approximately 10 percent of the occupiable habitat that existed at the time the HCP was prepared, and some 16 percent of the amount of unshielded, natural blowsand habitat (table 2).

The Coachella Valley Preserve in particular contains a fairly large and impressive remnant of desert ecosystem. In addition to providing habitat for the fringe-toed lizard, the preserve is home to 180 other animal species, among them an extensive resident and migratory bird population including burrowing owls, kestrels, and roadrunners. Some 120 different plant species are found on the preserve. It also harbors four other rare species: the flat-tailed horned lizard, the Coachella round-tailed ground squirrel, the giant red velvet mite, and the giant palm-boring beetle.

The land acquisition proposals contained in the HCP have been largely implemented. The vast majority of the acreage planned for acquisition under the plan has been acquired. The California Nature Conservancy took the lead in purchasing the core lands in the Coachella Valley Preserve, and had purchased options on the largest parcel (the Cathton Investments parcel) even before the HCP was completed. To date The Nature Conservancy has purchased 12,087 acres. A total of $18,201,365 was paid by TNC for land with a fair market value of $25,136,350. These figures reflect the fact that some lands were acquired through gifts or "bargain sales." Most of these lands have since been conveyed to the USFWS, the BLM, or the State of California, and TNC continues to own just 880 acres, the Thousand Palms oasis (Johnson 1989). Only an estimated 500 acres remains to be purchased in the Coachella Valley Preserve.

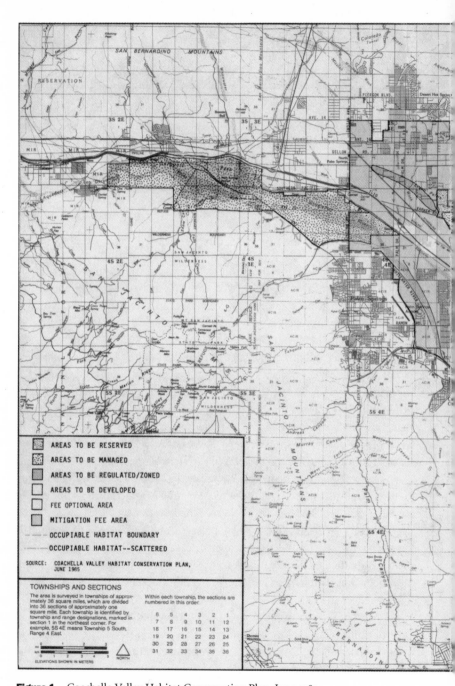

Figure 1   Coachella Valley Habitat Conservation Plan, June 1985

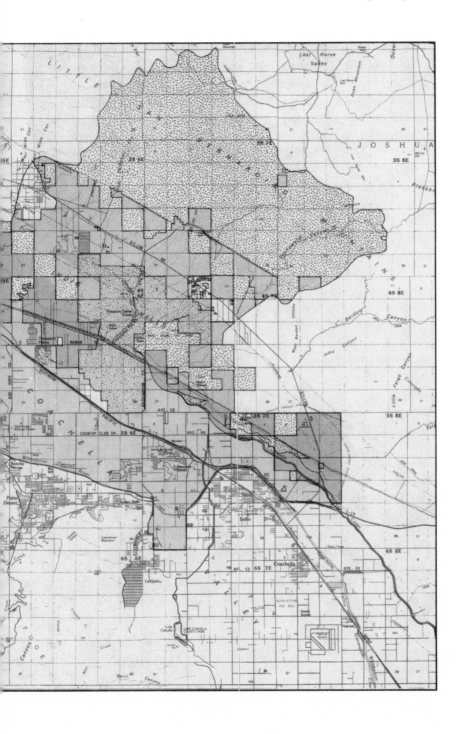

The other two preserves have been established as well, but none of the additional private land planned for acquisition at the Willow Hole-Edom Hill Preserve (nearly 800 acres) has been secured. The BLM still intends to acquire these lands when sufficient funds become available.

While some land would be protected through fee-simple acquisition, the HCP envisioned that certain other public lands would be managed so as to protect lizard habitat. When these "managed areas" are taken into consideration, the total area of conserved habitat jumps to approximately 12,300 acres. This is about 15 percent of the remaining occupiable habitat, and about 26 percent of the remaining unshielded habitat (Coachella Valley HCP Steering Committee 1985, v–13.).

The HCP also envisioned that certain privately held lands, specifically those near and around the Coachella Valley Preserve, would be subject to land use controls that might provide additional protection for lizard habitat. Specifically, much of the land in the area of the Coachella Valley Preserve is zoned "Natural Assets". On lands that fall under this zoning category, only one single-family dwelling unit is permitted by right per twenty acres. It was felt that such a low-intensity development pattern would generally help to protect habitat values and to minimize impact on the blowsand transport system.

The total projected cost of establishing the preserves was $25 million, to be obtained from a combination of sources (table 3). The cost of acquiring land for the Coachella Valley Preserve was estimated at approximately $20 million, with approximately $2 million needed to acquire supplemental land for the Willow Hole-Edom Hill Preserve. In addition to moneys to cover the costs of acquisition, it was assumed that funds would be needed to maintain the preserves. The HCP estimated these reserve maintenance costs to be between $50,000 and $125,000 annually. (Coachella Valley HCP Steering Committee 1985.) The HCP anticipates that these annual reserve maintenance costs will be generated by a special trust fund established toward the end of the acquisition program, the principal of which will be approximately $2.5 million.

Federal funding in the amount of $9.6 million from the Land and Water Conservation Fund has been allocated to the project, as planned. The Nature Conservancy has also met its $2 million fundraising goal, again according to plan. The most serious problem with the plan's funding has been a significant shortfall in the revenue generated from mitigation fees. As of 30 June 1989, $2,881,994 had been collected through the county and the nine cities. While the HCP projected annual revenues from development mitigation fees to be approximately $864,000, the amount collected has averaged only about $600,000 per year. The shortfall has been partly blamed on a depressed second-home market, the result of changes stemming from the 1985 Tax Reform Act (specifically, the suspending of mortgage

Table 3   Projected Distribution of Funding for Coachella
Valley Fringe-Toed Lizard Preserves

Funding Sources

Federal	
Land and Water Conservation Fund	$10 million
BLM land exchange (cash value)	5
State Wildlife Conservation Board	1
Nature Conservancy	2
Developer mitigation fees	7
	$25 million

Source: Coachella Valley HCP Steering Committee (1985).

deductions for second homes). The market has picked up in the last several years, in part because of the restoration of these second-home subsidies. A second explanation of the fee shortfall has been that several of the development "hot spots" in the valley have actually been outside of the mitigation fee zone (e.g., areas in Palm Springs that were already very attractive for development).

Once the habitat conservation plan had been completed, the local governments of Coachella Valley jointly became the applicants to the U.S. Fish and Wildlife Service for a Section 10(a) "incidental take permit." In April 1986, the USFWS issued the 10(a) permit for a period of thirty years. The permit is revocable at any point during this period, should USFWS feel the plan is not being followed. The HCP is implemented through two sets of legal agreements, executed concurrently with the issuance of the Section 10(a) permit. The first agreement is between the local governments, the U.S. Fish and Wildlife Service, and The Nature Conservancy. It sets forth the obligations of local governments to collect the mitigation fees and the obligations of The Nature Conservancy to manage and further acquire preserve lands. The second agreement is a preserve management agreement, necessitated by the multiple ownership of the Coachella Preserve, between TNC, the BLM, the California Department of Fish and Game, and the USFWS.

Evaluating the Coachella Valley Experience

As the number of federally listed endangered species continues to rise, and as development continues to extend into important habitat areas, these types of conflicts concerning conservation and development will undoubtedly increase in number. There are now more than fifty different habitat conservation plans either completed or in progress (most are still in preparation). The HCP mechanism will likely receive even greater use in the future (for a review of these different

HCPs, see Beatley 1991). Given these trends, the Coachella Valley HCP, as the second plan to be completed and approved by the USFWS, provides an opportunity to consider the effectiveness and desirability of this device, and to identify any major issues of implementation and policy that may have emerged from this early experience.

The accomplishments of the Coachella Valley habitat conservation plan are impressive, to say the least. Setting aside preserves of eventually close to seventeen thousand acres is a significant environmental accomplishment. The Coachella case, moreover, illustrates the tremendous benefits of constructive compromise between otherwise warring factions. Together, environmental and development interests accomplished what individually would have been difficult. Virtually every participant interviewed for this study regards the outcome as a "win-win" solution. There are few detractors, and most people seem to have gained feelings of pride and proprietorship about the plan. Indeed, the plan has received considerable publicity as an ideal model for resolving future conservation-development conflicts.

For the environmental community, the process allowed the marshalling of the financial and political resources to establish a preserve system and to protect large areas of blowsand habitat that for years prior to 1983 were lacking. The HCP provided a focal point and mechanism for garnering federal and state monies, and for assessing the private sector (through mitigation fees) for damages to habitat.

From the point of view of the development community, it was highly desirable to find a solution to the endangered species problem. Particularly in the cases of pending projects, extensive delays would have been costly. Developers realized that they would have to confront the lizard issue for each subsequent project, and saw the clear financial advantage of solving the problem "once and for all." Furthermore, finding a comprehensive long-term solution would be much more efficient than preparing individual biological assessments for each new project and engaging in a protracted battle over each new proposal.

The results of the Coachella Valley experience are not all positive, however; they illustrate the many practical economic and political difficulties of protecting endangered species and biodiversity in the context of rapidly urbanizing areas. In addition, a number of fairly basic policy questions, many philosophical in nature, can be expected to emerge in these efforts, as illustrated by the Coachella Valley case.

The ultimate test of whether any habitat conservation plan is successful, at least in the eyes of regulatory agencies such as the U.S. Fish and Wildlife Service, ought to be whether the lizard's survival is assured. Under the Endangered Species Act a Section 10(a) permit is to be issued only where it can be shown that the taking "will not appreciably reduce the likelihood of the survival and recovery of the species in the wild" (Endangered Species Act, Section 10[a][2][B][iv]). Most of the participants interviewed for this study admitted that the HCP was a gamble and that there was no assurance that the lizard would survive in the long term. The HCP itself acknowledges that the understanding of the biology of the lizard is quite limited. Particularly troubling are the dramatic population swings exhibited in the lizard population. Since 1986, an annual lizard count conducted by TNC confirms that lizard populations are prone to "wild yearly fluctuations" (Barrows 1992; Muth 1989). However, there is a general consensus that lizard populations appear to decline following years of low rainfall, as a result of the reduced amounts of vegetation and consequently diminished sources of food. Lizard counts in the last two years suggest that numbers have been on the rise (at least in two of the reserves), again largely corresponding to increases in precipitation (Barrows 1992).

Interviews with members of the scientific community disclose little confidence that the lizard's continued existence is assured. Most are quick to observe, however, that the chances of survival with the preserves are much greater than they would be without them. This highlights a general problem that arises in development-conservation conflicts. While years of careful scientific research on a species and its habitat may ideally be needed to understand fully its conservation needs, the time frame of the development community is much shorter. Its desires are to move ahead rapidly—and where problems such as endangered species arise, to overcome the problems quickly. These two perspectives are frequently in conflict.

Some have strongly criticized the Coachella Valley scheme as one that protects a relatively small portion of native habitat while opening up vast portions of the valley for development. Though close to seventeen thousand acres will eventually be acquired through the three preserves, they include only about seventy-eight hundred acres of occupiable lizard habitat. One recent commentator has argued that the Coachella Valley HCP "can only be described as doing, under the guise of an 'incidental take,' great violence to the concepts of recovery and survival" (Webster 1987).

Others object to "giving up" on existing lizard populations in areas outside of the preserves. Several policy questions arise here. While there has certainly been degradation of habitat due to sand shielding, the sand transport system is poorly understood, and the actual extent of degradation due to sand shielding is debatable. The Coachella Valley case illustrates the general uncertainty that exists concerning what levels of stringency in conservation and mitigation are required in HCPS, and the USFWS has been able to give little tangible guidance as to what is and is not acceptable.

Questions have been raised about the boundaries and design of the preserve system, highlighting the biological significance of these practical and political decisions. The biological integrity of the Coachella Valley Preserve, for example, has been questioned by some because of its failure to encompass a large portion of its designated critical habitat. The configuration of the preserve might have been more east-west, rather than north-sourth, to better reflect wind patterns that contribute sand from areas west of the preserve. Acquisition of the western land would have been difficult, however, because of the heavy parcelization of land that had occurred there, a situation illustrating some of the practical problems of assembling an "ideal" preserve system in a rapidly urbanizing environment.

The Coachella Valley case illustrates as well the general problem of attempting to set aside habitat "patches" in what is, or will become, an urbanized environment. The Coachella Valley preserve system is vulnerable to adjacent activities, and the long-term survival of the lizard may ultimately depend on the ability to control these uses. This problem has been highlighted in recent months in the Coachella Valley as a result of a proposal to build an auto racetrack in close proximity to the Whitewater River satellite preserve. Lizard counts in this preserve have been very low, and the site appears to suffer from the lack of sufficient sand entering the site. The sand transport system is, again, poorly understood but may be seriously jeopardized by activities occurring outside the preserve.

The high cost of habitat conservation is perhaps the most serious obstacle to the effectiveness of HCPs. This presents an interesting paradox. On the one hand, if society waits too long to purchase critical habitat areas, the public cost may be exorbitant. The greater the development pressures become, the higher the market values of these lands. On the other hand, it would have been difficult to have secured public and political support to buy such habitat areas in Coachella Valley far in advance—say fifty years ago. Indeed, there would have been no perceived "problem," and no officially listed endangered species. It is only because of the growth and development in the valley that the species is threatened to begin with. Thus, a vicious circle develops where efforts to set aside habitat preserves, as in the Coachella Valley case, must generally await development pressures, yet if conservation and public agencies wait too long, it may be too expensive to protect enough land to ensure survival of the species.

An important aspect of the funding problem is the general question of who ultimately ought to pay for the costs of preserving endangered species. There is considerable disagreement on this distributional question. Many participants in the Coachella Valley HCP process seemed comfortable with the position that because the plan was necessitated by a federal mandate (i.e., the federal Endangered Species Act) the costs should be shared by the broader national citizenry. A major point of disagreement, as we have seen, was the amount of contribution required of developers. The political and practical feasibility of higher developer fees aside, it could be argued that the development community got off rather cheaply. Its total contribution will amount to approximately $7 million of the total $25 million, or about 28 percent. This could be considered relatively modest, given that it is the activities of these developers that are endangering the species in the first place.

Furthermore, it is unlikely that a federal contribution such as that provided in the Coachella Valley case will be possible in many future habitat conservation plans. With more than 750 species currently listed as either endangered or threatened, it seems neither desirable nor possible to expect the federal government to contribute consistently as much as 60 percent of the costs of necessary management and protective programs.

There is also considerable ethical appeal in the position that for every acre of habitat destroyed or disturbed by development, an acre should be protected somewhere else. The area of environmental mitigation offers substantial precedent for such a compensation ratio. For instance, an acre-for-acre mitigation standard is frequently imposed under federal and state wetlands laws, and in some states the replacement ratio is 2:1 or higher (Kusler 1983). For public projects that must go through state consultations under California's Endangered Species Act, the California Department of Fish and Game is now typically requiring mitigation on the order of 3:1. On the basis of market land prices of about $3,500 per acre (now very conservative), Coachella Valley developers are being required to replace each lost acre with perhaps 0.2 acres of protected habitat. Moreover, as land prices continue to escalate and the mitigation fee remains constant, the effective mitigation ratio tends to go down over time.

The Coachella Valley case raises fundamental questions about the extent to which we can preserve all endangered species. While the Coachella Valley habitat conservation plan may go far in ensuring the continued existence of the fringe-toed lizard, critics have questioned the logic of expending large amounts of money and social resources—in this case some $25 million—to achieve such an objective. Some have suggested that such funds could be better used to combat poverty in the barrios of Los Angeles, or to meet some equally worthy social need. Although they understand the usual arguments for preserving species (e.g., pharmaceutical benefits), they are still skeptical about the amount of societal return from such an investment. To some, this feeling was heightened by the close resemblance of the Coachella Valley fringe-toed lizard to two related fringe-toed lizards indigenous to the United States (the Colorado Desert and Mojave Desert fringe-toed lizards).

The Coachella Valley habitat conservation plan also raises questions about the merits of focusing on a single species. While the preserves created under the HCP will protect habitat for numerous other species, the specific biological focus on the fringe-toed lizard is somewhat misguided. It is increasingly argued that HCPs must take a broader, multispecies or ecosystem approach, and that such an approach makes better sense both from a biological and an economic perspective.

Even where concern about a single endangered species may be justified or appropriate, focusing on the larger ecosystem may bring great political benefits, as the fringe-toed lizard case illustrates. Throughout the history of the lizard conflict, most elected officials, developers, and average constituents did not generally support actions—particularly expensive actions—to preserve the lizard. Had the species been a more visually attractive or symbolically important one, such as a bald eagle or a Florida panther, the concern might have been greater. This reaction is consistent with research showing that people tend to attach greater importance to protecting the larger, more attractive species, and less importance on species like snakes, insects, and plants (Kellert 1979). Rather, what was ultimately acceptable and justifiable in the eyes of Coachella Valley officials and constituents was the setting aside of a segment of the natural desert environment to be enjoyed by future generations.

Finally, when looking at the Coachella Valley approach one is struck by a failure to question or address in any fundamental way the engine driving habitat loss. The Coachella Valley plan, like most others, does not challenge development trends but rather accepts them, seeking to find a conservation solution that will allow these growth patterns to continue. Yet, part of the logical long-term solution might be to reconsider seriously and fundamentally the current development patterns and to make adjustments to allowable densities and uses, and so reduce the overall extent of habitat loss. Need development projects be as large or as sprawling as they are? Can units be smaller and development more compact and contiguous? Can recreational amenities such as golf courses be scaled back to better conserve limited land? These types of land use and growth assumptions were not, however, explicitly addressed in the Coachella Valley habitat conservation plan.

In conclusion, the Coachella Valley HCP is certainly impressive in its accomplishments. It makes clear the great benefits that can result from compromise and cooperation between typically warring interest groups. Moreover, the amount of land set aside and protected as a result of the plan and the generation of funds and resources necessary to bring this about are equally impressive. The Coachella Valley experience, however, is not without certain limitations, some of which have been identified here. Furthermore, regardless of how successful one feels the Coachella Valley HCP has been or will be, it illustrates effectively the many practical obstacles—political, economic, biological—that must be overcome in this type of habitat conservation planning.

Acknowledgments

Funding for this article was provided by the National Fish and Wildlife Foundation. Any opinions, findings, conclusions, or recommendations are those of the author and do not necessarily represent the views of the foundation. This article is adapted from an article entitled "Balancing Urban Development and Endangered Species: The Coachella Valley Habitat Conservation Plan," published in *Environmental Management* vol. 15, no. 6, (1991) and used by permission of Springer-Verlag Publishers.

References

Barrows, C. 1992. *Monitoring report: Coachella Valley fringe-toed lizard.* Southern California Area Manager, The Nature Conservancy. Thousand Palms, CA.

Beatley, T. 1990. *Land development and protection of endangered species: A case study of the Coachella Valley habitat conservation plan.* Prepared for the National Fish and Wildlife Foundation. Washington.

———. 1991. Use of habitat conservation plans under the federal Endangered Species Act. In *Wildlife conservation in metropolitan environments,* ed. L. W. Adams and D. Leedy. Columbia, MD: National Institute for Urban Wildlife.

Carpenter, C. C. 1963. Patterns of behavior in the forms of the fringe-toed lizards. *Copeia* 2:406–12.

Coachella Valley Association of Governments (CVAG). 1988. *Coachella Valley area growth monitor.* Palm Desert, CA: CVAG.

———. 1989. *Regional housing needs assessment.* Palm Desert, CA: CVAG.

Coachella Valley HCP (habitat conservation plan) Steering Committee. 1985. *Coachella Valley habitat conservation plan.* Riverside, CA.

Cornett, J. 1983. Uma, the fringe-toed lizard. *Pacific Discovery* 36 (April–June): 1–10.

Defenders of Wildlife. 1987. *Saving endangered species: Implementation of the Endangered Species Act.* Washington: Defenders of Wildlife.

England, A. S., and S. G. Nelson. 1976. Status of California Department of Fish and Game, The Coachella Valley fringe-toed lizard. Inland Fisheries administrative report no. 77-1. Sacramento, CA.

General Accounting Office. 1988. *Endangered species: Management improvement could enhance recovery program.* Report GAO/RCED-89-5, December. Washington: General Accounting Office.

Johnson, J. 1989. The California Nature Conservancy. Personal communication.

Kellert, S. 1979. *Public attitudes toward critical wildlife and natural habitat issues.* Washington: U.S. Fish and Wildlife Service.

Kusler, J. 1983. *Our national wetlands heritage: A protection guidebook.* Washington: Environmental Law Institute.

Marsh, L. L., and R. D. Thornton. 1987. San Bruno Mountain habitat conservation plan. In *Managing land use conflicts: Case studies in special area management,* ed. D. J. Brower and D. S. Carol. Durham, NC: Duke University Press.

Moore, S. 1983. Fringe-toed lizard flares anew in desert area. *The Press-Enterprise.* 2 July.

Muth, A. 1989. Population biology of the CVFTL. Progress report number 3. Submitted to California Department of Fish and Game. Sacramento, CA.

Reid, T. S., and D. D. Murphy. 1986. The endangered mission blue butterfly. In *The management of viable populations: Theory, applications and case studies.* ed. B. Wilcox, P. Brussard and B. Marcot. Palo Alto: Stanford University, Center for Conservation Biology.

Riverside County Planning Department. 1989. Interim habitat conservation plan for the Stephens kangaroo rat. Prepared by regional environmental consultants. Riverside, CA.

Schlaepfer, G. 1985. A model for land conservation: An oral history analysis of habitat preservation for the Coachella Valley fringe-toed lizard, an endangered species. Master's thesis, California State University, Fullerton.

The Press-Enterprise. 1983a. Rancho Mirage to sue over lizard. 19 November.

———. 1983b. Rancho Mirage favors desert preserve. 3 October.

———. 1983c. County planners tentatively favor condos although lizards are issue. 24 June.

———. 1984. CVWD to add 1200 acres to preserve. 25 May.

U.S. Fish and Wildlife Service. 1984. Recovery plan for the Coachella Valley fringe-toed lizard. Portland: USFWS.

Webster, R. W. 1987. Habitat conservation plans under the Endangered Species Act. *San Diego Law Review* 24:243–71.

Yaffee, S. L. 1986. *Prohibitive policy: Implementing the federal Endangered Species Act.* Cambridge, MA: MIT Press.

The Metropolitan Portland Urban Natural Resource Program

Joseph Poracsky and Michael C. Houck

Introduction

Situated at the confluence of the Columbia and Willamette rivers, the Portland-Vancouver metropolitan region encompasses parts of four counties in northwestern Oregon and southwestern Washington (fig. 1). Portland, with a population of 437,289, is by far the largest city in the region, including much of Multnomah County and encircled by rapidly urbanizing portions of Clackamas and Washington counties. Immediately to the north of Portland on the Columbia River is Clark County and the city of Vancouver, the major city in southwestern Washington, with a population of 46,380 (U.S. Bureau of the Census 1991b). To the east of Portland is Gresham, Oregon's fourth-largest city, with a population of 68,235. To the west are Beaverton and Hillsboro, the fifth- and ninth-largest cities in the state, with populations of 53,310 and 37,520 (U.S. Bureau of the Census 1991a). The population of the metropolitan region, which includes thirty-one cities, totals 1.24 million people and in the next twenty years is projected to grow by about 500,000.

Most of the region's lower elevations are in either agricultural or urban uses. Although the originally coniferous-forested higher elevations have been partially developed or harvested for timber, large portions remain in second-growth forest, generally conifer and mixed conifer-deciduous. The region's climate is a modified West Coast marine climate, with mild, wet winters and clear, dry summers (Johnson 1987, 21) that provide an hospitable environment for a diverse array of plant and animal communities.

Urbanization has greatly altered the ecological character of the region. In addition to the superimposing of a dense network of development and built structures, other actions such as the removal of native vegetation for agriculture and timber, modification of stream channels, filling and draining of wetlands, and the introduction of nonnative species of vegetation and wildlife have all contributed to the modification of the landscape.

Despite these alterations, however, development has not gone out of control, and the region retains a lush, green appearance and a relatively intact system of

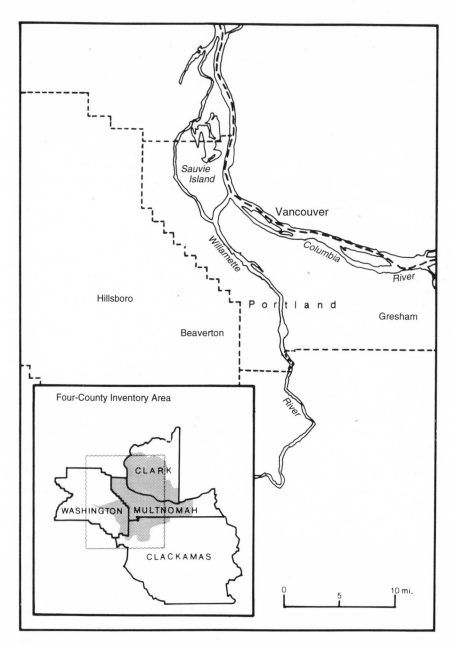

Figure 1   The Portland-Vancouver metropolitan region and key places identified in the text. The shaded portion of the inset map indicates the area included in the four-county inventory.

natural areas. In part, the natural appearance is due to the fact that the region is rimmed by scenic views of evergreen-clad mountains surrounding the valley and of snow-clad Mount Hood to the east and Mount Saint Helens to the north. However, the large number of unique and highly visible natural areas within the urbanized metropolitan region itself is a more important factor in helping to maintain the region's sense of livability.

Foremost among the region's natural areas is Forest Park, the largest city park in the United States. Straddling the Tualatin Mountains on the northwestern edge of Portland, this five thousand-acre area provides fifty miles of forested trails within a few minutes' drive from downtown (Houle 1988). Other large, well-known natural features include Sauvie Island, Vancouver Lake lowlands, Ridgefield National Wildlife Refuge, Salmon Creek Greenway, the Columbia Slough, Smith and Bybee lakes, the Sandy River Gorge, Oaks Bottom Wildlife Refuge, Powell Butte Nature Park, Johnson Creek, the Willamette River Greenway, the Tualatin River, Fanno Creek, and forested areas along Skyline Drive, north of Forest Park.

There are also many smaller and less-well-known areas that are equally important from a local perspective. Among these are Elk Rock Island, Heron Pointe Wetlands, Camassia Nature Conservancy Preserve, Reed College Canyon, Beggars Tick Marsh and Oregon Episcopal School Marsh.

The natural landscape features not only represent important visual amenities, but they also provide habitat for a variety of wildlife not usually associated with a large city. For example, Forest Park is home to elk (*Cervus canadensis*) and an occasional black bear (*Euarctos americanus*). The park's Balch Creek watershed contains over three thousand native cutthroat trout. Wetlands and riparian forests on Sauvie Island and around the Vancouver Lake lowlands attract bald eagles (*Haliaeetus leucocephalus*) that feed on winter waterfowl, making the daily migration from roosts in old-growth forests to wetlands west of the metropolitan region. These and other wetlands also attract great blue herons (*Ardea herodias*), sandhill cranes (*Grus canadensis*), Canada geese (*Branta canadensis*), and a variety of other waterfowl.

In sparsely developed areas on the edges of the metropolitan region are found coyote (*Canis latrans*). Most metropolitan waterways are populated with beaver (*Castor canadensis*), and many streams still maintain the annual run of several species of salmon, such as the chinook (*Onchorhynchus tshawytscha*).

Both the diversity and interspersion of habitats continue to support a rich fauna within the region. However, the existing variety of local natural areas is the result of a fortuitous combination of economic conditions, local topographic conditions, happenstance, and occasional dogged determination by individuals and grassroots organizations. As the metropolitan region continues to gain population and grow, the future status of wildlife and their habitats comes into question. As the number and quality of habitats decline, the overall quantity and variety of wildlife will also decline. The single most important cause for the reduction of biodiversity is the "destruction of habitats by conversion of natural landscapes to more intensive uses" (Jenkins 1976, 44).

Of the great number of ecologically rich areas remaining in the metropolitan region, only a small portion enjoy protection. Most natural areas, because they are privately owned, are available for development and are increasingly threatened by the chain saw and the bulldozer. Although limited protection is afforded selected individual sites throughout the state under Oregon's land use program (Land Conservation and Development Commission 1985), there are no local programs that address the protection or management of urban natural areas from an eco-system perspective (Rogers and Houck 1984).

Along with an increasing loss of natural areas has come the accelerated con-version of agricultural lands to urban uses. Many of the farms of the metropolitan region represent links to pioneer history and the permanent settlement of the region by nonindigenous people. These farms have important cultural connec-tions to the old "Oregon Country" and provide an historical "sense of place." In addition, agricultural areas comprise a "working landscape" (Hiss 1990, 114–15) that contributes a multitude of benefits including visual diversity, a sense of agrar-ian or rural tranquillity, seasonal wildlife habitat, and economic value. The amount of working landscape being converted to urban uses is great. One study, describing changes in Clackamas and Washington counties in the 1970s, indicated that "more than two-thirds of the land that shifted to urban residential use came from cropland" (Vesterby 1987, 158).

The pace and visibility of the losses in agricultural land have aroused a public perception that more than just farmland is being destroyed, and that a part of the local history and character is disappearing. Many residents throughout the region value both the natural areas and working landscapes very highly, a point that was emphasized in a recent series of public workshops (Metro 1991c). Increasingly, citizens are concerned that the region's unique identity is being taken away and that the livability of the metropolitan area is being compromised.

In response to this concern, a major effort in landscape protection and enhance-ment is underway in the Portland-Vancouver metropolitan region. The "Metro-politan Greenspaces Program" represents a unique, coordinated initiative to iden-tify, protect, and manage landscape, wildlife, and the sense of place within a rapidly growing metropolitan region. The evolution of the program and its goals, and the program's early successes provide an example that may contribute useful insights to similar efforts in other metropolitan areas.

Genesis of Metropolitan Greenspaces Program

Numerous sources have contributed to the development of regional approaches to landscape and natural area preservation. For the Portland-Vancouver metropolitan region, relevant contributions have been traced back to such well-known figures as John Charles Olmsted in 1903 and Lewis Mumford in 1939 (Houck 1991). However, the genesis of the Metropolitan Greenspaces Program can be traced more directly to three recent local initiatives, all of which have drawn their inspiration from the works of Olmsted, Mumford, and others but have evolved separately from them.

The first initiative was the incorporation of the Forty-Mile Loop Land Trust in 1981 (Edelman 1991). In both name and purpose, the trust is derived from a proposal by John Charles Olmsted in a 1903 study for the Portland Park Board. His proposed "comprehensive" system of parks, parkways, and boulevards formed a circuit around turn-of-the-century Portland (Olmsted Brothers 1903). The modern-day Forty-Mile Loop, which now comprises over 140 miles of bicycle and pedestrian trails, is recognized as "one of the most creative and successful greenway projects in the country" (Little 1990, 77).

The second initiative was a proposal to establish a Metropolitan Wildlife Refuge System. The goal of the refuge system, launched by the Audubon Society of Portland, is to create a system of interconnected natural areas that would be managed primarily as wildlife habitat and wildlife viewing sites. In a fashion similar to that of the Forty-Mile Loop Land Trust, the refuge system is intended "to provide an appropriate organization for the donation of significant lands and conservation easements" and to promote management, appropriate public use, and scientific research on natural areas (Lev and Houck 1988, 25). Though the refuge system is not yet a physical reality, the concept has developed into a locally recognized program that has obtained both local financial backing and an increasing amount of public support.

The third initiative was a regional parks inventory by the Metropolitan Service District (Metro), a regional government having jurisdiction within Multnomah, Washington, and Clackamas counties, the three urbanized Oregon counties constituting the metropolitan region of Portland. Metro's 1989 study dealt with a broad spectrum of recreational spaces, including ball fields, swimming pools, tennis courts, picnic areas, and nature parks. One of the needs that study identified was for "a comprehensive information base" that would include the "cataloging and evaluation of existing natural areas throughout the metropolitan area" (Metro 1989, 37).

Soon after completion of the Metro study, it was recognized that substantial functional and geographical overlap existed among the Forty-Mile Loop Trail program, the proposed Metropolitan Wildlife Refuge System, and the natural areas identified in the Metro regional parks study. The energies of the individual organizations associated with these three initiatives quickly coalesced into an alliance of interests that was shortly joined by the government of Clark County, Washington, which had just begun to address a similar set of concerns through its fledgling Open Space Commission. The result was the initiation of an inventory and analysis of natural areas designed to weave the numerous overlapping strands of interest into a single net covering the four-county, bistate metropolitan region.

Metro, the agency with planning authority over the largest portion of the metropolitan region, took the lead in coordinating the multijurisdictional effort to inventory natural areas. This bistate cooperative effort has involved nearly forty local and regional jurisdictions and several state and federal sponsors (Metro 1991a).

The Natural Area Inventory

The short-term goal of the inventory was the compilation of a data base of natural areas, including the preparation of a map from aerial photographs and the collection of detailed field data for a small sample of sites. Planning for the inventory began in February 1989, and the inventory was completed in December 1990 (Poracsky 1991). The inventory encompassed 918 square miles, or 25 percent of the 3,653-square mile four-county area. The inventory was performed on contract by the geography department at Portland State University and two wildlife consultants, Lynn Sharp and Esther Lev. All aspects of the inventory were designed and reviewed by a technical advisory committee of local biologists and aerial reconnaissance specialists and by the Metro Parks and Natural Areas Forum, a loose-knit consortium of more than eighty elected officials and their staffs, planners, and environmental and public advocacy groups (Poracsky, Sharp, and Lev 1991).

The inventory involved four steps. The first was to acquire current aerial photographs as the data source for the compilation of a map depicting natural areas. In addition to its scientific uses, the acquisition of aerial photographs proved to be an important mechanism for bringing a broad coalition of interests into the project. In return for contributing to the cost of the $20,000 flight and the $89,000 inventory and data base development effort, cooperators were offered a reduced "sponsor's rate" when purchasing aerial photos.

The second step utilized color infrared aerial photographs to identify and map natural areas. Informally, a natural area may be viewed as a self-sustaining area that would not change dramatically if all human influences were removed. Formally, for the study, a natural area was defined as "a landscape unit (a) composed of plant and animal communities, water bodies, soil, and rock, (b) largely devoid of man-made structures, and (c) maintained and managed in such a way as to promote or enhance populations of wildlife" (Poracsky, Sharp, and Lev 1991). This definition served not only to describe what would be included, but also to clarify which landscapes would be excluded from the study. The two most prominent examples are golf courses and agricultural land, both of which were eliminated on the basis that they are not normally managed for wildlife habitat values. Despite the decision to exclude them, it was recognized from the earliest stages of the study that golf courses and farmland may play a critical role in providing wildlife habitats and must eventually be included in a truly comprehensive inventory. Nonetheless, their elimination through the definition was intentional, in order to limit the budget of the study. It is anticipated that data concerning these two classes of features will be obtained as part of a set of land use data and will figure prominently in later analysis of the natural areas data.

The characteristics identified in the definition of natural areas were incorporated into a detailed classification system utilized during the process of photo interpretation and mapping. Through the use of color infrared aerial photographs, each natural area was delineated on a map and annotated with a code that identified

(1) whether it was an upland or a wetland site;
(2) the cover category, i.e., forest (with three subcategories based on canopy closure), scrub-shrub (also with three subcategories), clear cut, meadow, bare soil, or rock;
(3) the percentage of woody vegetation in deciduous cover (estimated in 10 percent increments) for the forest and scrub-shrub categories;
(4) whether the site was riparian, i.e., adjacent to water.

Maps were derived from the visual interpretation of 1:24,000-scale enlargements of 1:31,680-(nominal) scale color infrared aerial photographs. Adjustment of individual photo scales was necessary in order to correct for local changes in elevation and to obtain a common scale on all the photos. The 1:24,000 scale was employed because it was a convenient match for a readily available local digital base map. Approximately two hundred fifty photographs were employed and almost sixty-four hundred distinct natural areas were mapped, varying in size from about one acre to nearly one thousand acres. Identification of each natural area was based solely on the cover characteristics and the land use of the site as interpreted from the photographs. The question of whether the land was publicly or privately owned was not considered, though that data will be acquired and studied in subsequent analysis.

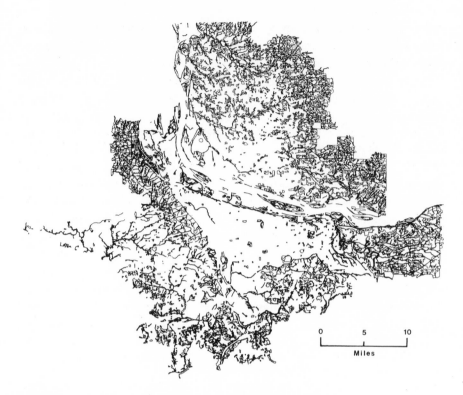

Figure 2   Natural areas in the Portland-Vancouver metropolitan region

The third step in the process involved collecting on-site data for a sample of the mapped areas. Since the mapping produced very limited, broad-scale information, field surveys were needed to obtain more detailed data. A sample representing a cross section of the categories of mapped natural areas was selected, and 157 sites were visited by a team of field biologists. They collected a variety of on-site data, including the species of plants and animals, the identification of wetland types, the presence of trails and other human disturbances, and the overall habitat quality (Lev and Sharp 1991).

The fourth step was entering map and field survey data into a spatial data base. First, the photo-interpreted natural areas were digitized to produce a data layer within a geographic information system (GIS). Second, another data layer was prepared by entering the detailed field data into the GIS and relating the data to the appropriate photo-interpreted sites.

The GIS provides flexibility for statistical and graphical outputs, and creates a number of alternatives for subsequent analysis. For example, figure 2, a portrayal of the natural areas that were mapped in the four-county metropolitan region, was produced from the GIS data base. Table 1 presents a summary of the mapped natural areas for the same four-county region.

Analysis of the Inventory Data

A primary objective of the Metropolitan Greenspaces Program is to ensure that sustainable plant and animal communities are incorporated into a protected system. The analysis of the inventory data contributes to that goal by establishing criteria for evaluating the ecological functions of, and connections between, natural areas. This evaluation in turn assists in developing a basis for assigning priorities to potential sites for acquisition in public and nonprofit programs.

The evaluation encompasses the human and wildlife values as well as the ecological viability of natural areas. Among the various human values, an important one is access. A good example of this value is provided by Forest Park, with its fifty miles of trails in nearly five thousand acres. Ready access for hikers, nature watchers, joggers, and mountain bikes (on selected trails) ensures Forest Park a major role in the recreational patterns of many residents of the metropolitan region. One portion of the analysis considered the distance from residential populations as an evaluation factor. Other relevant human values include contributions to water and air quality, educational potential, and aesthetics.

Among wildlife values, foremost is the uniqueness of the site. For example, the native cutthroat trout population of Balch Creek represents a distinctive element that has been instrumental in achieving the successful protection of the stream. Without this unusual wildlife presence, the Balch Creek area would today be undergoing heavy residential development. A key part of the analysis resulted in recommendations on protecting distinctive and unique areas.

Also of concern are "common" sites, which any system of protection for the existing environment must include in order to reflect a comprehensive set of wildlife values. Such sites are important not only as representatives of habitats that

Table 1 Summary statistics for natural areas in the four-county inventory area of the Portland-Vancouver metropolitan region

Unit	Total Area (acres)	Natural Areas (acres)	% of Unit
Clackamas County	76,992	17,524	22.8%
Washington County	76,605	10,930	14.3%
Multnomah County	219,093	80,911	36.9%
Clark County	249,600	93,336	37.4%
Four Counties	622,290	202,701	32.6%

are common now, but also as ones with the potential to become unique in the future. Sites representing common local habitats must be valued and incorporated into a regional system of natural areas because changes in land use patterns may result in their becoming increasingly rare through time.

Another wildlife value is the interconnectedness of sites, which is an ecological necessity. As fragmentation of natural areas increases, it is accompanied by a reduction in the stability of the wildlife community (Curtis 1956, 734). Providing connections via natural corridors allows interactions between larger areas to occur with minimal impact on development (Diamond 1975, 144). The analysis focused particularly on stream and ridgeline corridors as potential connecting links.

While specific human and wildlife values may be dealt with separately, the significant element is the interaction between these values. An example of this type of interaction is found in the heron rookery adjacent to Heron Lakes Golf Course. Human disturbance of the rookery would be likely to result in nest abandonment. However, maintenance of a dense barrier of shrubs, which discourages physical access while providing visual access, has successfully protected the heronry. In this case, a simple but judicious management decision has ensured that the wildlife value of the site is preserved while the human value is enhanced and protected.

The analysis relied heavily on the use of geographic information system technology to relate the natural areas pattern to other factors such as land uses, zoning, comprehensive plans, utility rights-of-way, National Wetlands Inventory data, and soil and slope information. For example, a data layer has been developed to portray areas that currently have varying levels of protection by virtue of their designation as parks or wildlife refuges, or ownership by local land trusts. By overlaying this data layer with the one for the natural areas, it is possible to determine which portions are already protected and constitute potential components of a larger interconnected system.

One analysis approach that has not been systematically employed yet, because of lack of data, is to examine the uses of adjacent lands, especially agricultural lands, that might be suitable for inclusion in the system. Many kinds of farming operations, while not intended to support wildlife, are generally benign to wildlife and can provide useful habitat, especially on a seasonal basis. In addition, these types of lands are potentially important as buffer zones between natural areas and high-density development and, when included as part of a total system, might offer enough protection to make an otherwise undersized or isolated natural area a valuable acquisition. On the other hand, some adjacent land uses might preclude the acquisition of an otherwise valuable natural-area site, because the proximity of the land use might contribute to the accelerated degradation or destruction of the habitat. Additional work is currently underway to develop appropriate data to assist in this kind of analysis.

The Metropolitan Greenspaces Program

Initially, the Metropolitan Greenspaces Program was identified as an "Urban Natural Areas Program," with a focus limited to the inventory and analysis of natural areas and wildlife habitat data. As the program has evolved, its goals have expanded to include other natural and cultural resources, while continuing to concentrate on interconnectedness and natural areas. These expanded goals have resulted in the adoption of the new program title, Metropolitan Greenspaces Program, a name chosen to reflect its regional focus and the multiple benefits of "Natural Areas, Open Space, Trails and Greenways, for Wildlife and People" (Metro 1991b).

The program is intended to focus on *acquisition* and will potentially involve a combination of protection strategies. A primary strategy, purchase in fee simple from willing sellers, has already been initiated. The Multnomah County Commission (a cooperating jurisdiction) has established a trust fund, which receives 50 percent of the proceeds from the sales of unrestricted county properties, for the acquisition and management of urban natural areas. Levies passed during the November 1990 elections in Gresham and Lake Oswego are being used to acquire open space in those jurisdictions.

Alternatively, private landowners may donate properties. Numerous examples of successful efforts of this kind exist throughout the country (Mantell, Harper, and Propst 1990; Little 1990). While The Wetlands Conservancy, a local land trust, has been actively soliciting land donations, their efforts need to be expanded throughout the region. The Forty-Mile Loop Land Trust has been especially effective in obtaining hiking and biking corridors.

Funding for the Metropolitan Greenspaces Program has come from a variety of sources. The initial data acquisition and inventory cost about $109,000, paid through allocations from Metro, local jurisdictions, and state agencies. The analysis, demonstration project, planning, and public education and outreach functions currently underway account for about $100,000 from these same sources, supplemented by a $30,000 grant from the U.S. Environmental Protection Agency and a congressional appropriation through the U.S. Fish and Wildlife Service for $567,000.

Both the analysis portions of the program (Poracsky et al. 1992) and an acquisition and management plan (Metro 1992) have been completed. With these elements in hand, the next step will be to present the voters with a regional bond measure to fund property acquisition.

Building Public Support and Public Involvement

There is good reason to believe that the public strongly supports the goals of the Metropolitan Greenspaces Program and the passage of a regional bond measure to fund acquisition. In a 1990 survey, 81 percent of the region's population indicated "support [for] a bond measure to acquire and protect urban natural areas to achieve long term environmental balance in the Portland metropolitan area" (Market Trends 1990). These results are consistent with survey responses in other West Coast communities and with the recent passage of local levies in Gresham and Lake Oswego for the acquisition and management of natural areas and wildlife habitat.

Efforts to expand and strengthen the existing base of support are ongoing. Special activities and events, the formation of a regionwide citizen advocacy group, and the production of informational materials have been successful in creating grassroots support for the Metropolitan Greenspaces Program.

In the areas of special activities and events, a number of very successful examples may be cited. Frequent presentations by out-of-town specialists on urban natural resources and greenspace issues help to bring in fresh ideas and stimulate local citizens. Conferences such as the annual "Country in the City" symposium, a "Fragile Lands" symposium, and a conference oriented to the business community on "Livability and Profitability" have helped promote discussion and educate the public, elected officials, and the business community about the ecological and economic rationale for the Greenspaces Program. Annual public celebrations of nature such as "Great Blue Heron Week," celebrating Portland's official city bird, and the "Salmon Festival," which attracts up to ten thousand participants over a two-day weekend, include numerous field outings to natural areas. Newly instituted in 1991, a series of over 100 informal "Endangered Spaces" tours provides a regular vehicle for acquainting citizens with individual natural-area sites.

The region has a long history of activism by citizens and local environmental organizations. A recently organized grassroots group, the Friends and Advocates of Urban Natural Areas (FAUNA), seeks to educate the public and to coordinate the activities of over sixty local "Friends" groups throughout the region. FAUNA's primary mission is to assist in the development of numerous local land-steward and greenspaces advocacy groups in the region through workshops, publications, presentations, and lectures.

A variety of publications and informational materials have been developed to provide the public with background about the Metropolitan Greenspaces Program. A colorful brochure and map describing the program were printed and made available throughout the region. Supporting materials such as Portland

Audubon's "Urban Stream" brochures focus on the importance of stream corridors to the Metropolitan Greenspaces Program. Audubon's publication *The Urban Naturalist* offers maps and natural history information on existing greenspaces in public ownership. The FAUNA newsletter contributes to the potential for public involvement by informing various local groups and individuals about the Greenspaces Program and how they can best support and interact with it.

Conclusions and Implications

Metropolitan Greenspaces is an ambitious program that will require long-term political and public commitment before the ultimate goal of a "cooperative regional system" is realized. None of the ideas in the program are new, and the program has aggressively borrowed ideas from successful models such as the East Bay Regional Parks District.

Despite the magnitude of the remaining tasks, the program can already point to some striking achievements. For example, as a result of the inventory the region has a comprehensive map of natural areas and has initiated a systematic biological field inventory and data base. The data have both current and future values. A natural-area land type that is biologically defined, as opposed to a traditional designation such as "vacant" or "undeveloped," represents an uncommon and generally overlooked approach to dealing with the urban environment. Too often, data concerning urban areas include only economic and social parameters; since systematic data concerning the natural environment and people-nature interactions are not generally available, these parameters are generally ignored or receive only passing or anecdotal treatment by planners. However, given the growing demands by the public for environmental planning and the growing ability of planners to employ multiple layers of data in geographic information systems, it is likely that natural areas data will become an increasingly common component in the planning process, providing an important tool for guiding difficult decisions regarding the balance between urban growth and the maintenance of the quality of life. This would appear to be especially true in the Portland-Vancouver region, where the concept of the natural areas data base has been embraced by local jurisdictions.

The inventory data are already being put to use. Several local jurisdictions have begun to utilize the natural areas data and the ecological concepts embodied in the Metropolitan Greenspaces Program to guide them in making local acquisitions and in planning for potential pieces of the regional system.

Perhaps the most important achievement of the Greenspaces Program so far has been the degree of cooperation that it has engendered and the willingness on the part of a number of individuals and agencies to make the program work. Jurisdictions on both the Washington and Oregon sides of the Columbia River recognized the logic of an inventory based on the ecological unity of the metropolitan region and joined in a common data-collection effort. The program has generated unprecedented cooperation among cities, counties, and other local jurisdictions and has involved the public in greater numbers than any previous planning effort. More than twenty-five local jurisdictions have already passed a resolution of support for the establishment of a regional system like Metropolitan Greenspaces (Metro 1990). Because of differences in the land use laws and funding mechanisms in Oregon and Washington, the analysis of the inventory data is formally being performed separately for the two states. However, a great deal of cooperation, idea sharing, and coordination is going on to ensure complementarity between the individual analysis efforts.

The importance of cooperative, regional planning to this effort cannot be overstated. Local planning cannot provide the protection necessary for regional natural resources, especially riparian corridors and wildlife habitats that cover extensive areas and cross multiple jurisdictions (Collins 1990).

It is the unusual alliance of a broad group of jurisdictions, public constituencies, and the business community, that will determine the success of the Metropolitan Greenspaces Program. Through the establishment of a regional system of greenspaces and the development of an acquisition and management program, the Portland-Vancouver area will preserve its diverse natural habitats, which will support wildlife and provide people with a variety of opportunities for passive recreation.

The specter of uncontrolled population growth and public concern about loss of livability have rekindled interest in the visions of John Charles Olmsted and Lewis Mumford. The renewed cooperative effort toward truly regional planning has the potential to ensure that their vision is translated into the reality of a network of green areas within the fabric of the urban and urbanizing environment. The Portland-Vancouver metropolitan region is working vigorously to implement the vision and to incorporate both wildlife and human concerns within a new kind of diverse and sustainable natural urban ecosystem, defined not by arbitrary political boundaries, but by a coherent ecological region.

Postscript

In the November 1992 general election, Ballot Measure 26–1, seeking $200 million in general obligation bonds to support acquisitions under the Metropolitan Greenspaces Program, was put before the voters. The measure lost by a vote of 44 percent to 56 percent.

The election was an unusual one, to say the least. In addition to the national race to select a new president, which always draws media attention, there was an extremely close U.S. Senate race between Bob Packwood and Les AuCoin and a hard-fought mayoral race, both of which were major media concerns during the campaign season. Most significant, however, was the effect of the extremely divisive contest over Ballot Measure 9, involving a proposed state constitutional amendment to abridge the civil rights of homosexuals. Though Measure 9 was defeated, it provoked an especially rancorous and heated battle that drew the lion's share of campaign dollars, campaign volunteer energy, and, most important, media attention. Throughout the campaign, the major issue on the airwaves, in the press, and in private conversations was Measure 9.

Though Ballot Measure 26–1 lost, another measure adopting a "home rule" charter for Metro did pass. Among other things, the charter measure authorized Metro to (1) become involved in the operation and management of parks, (2) raise revenues for park operation and maintenance, and (3) perform regional planning functions, including open space planning. The voters thus chose to pass the authorization but not the appropriation for Metropolitan Greenspaces.

Preelection polling indicated that among voters who were aware of the ballot measure for the Greenspaces Program, support was very high. The problem was very low voter recognition of the measure. When confronted with it in the voting booth, the voters saw it as simply another attempt by government to wring money from them. Thus, as with every other measure in the state that asked voters for financing, 26–1 was defeated, not on its merits, but just because it was a money item.

Despite losing on the bond measure in the election, Metro and the many other people involved in the program have resolved to make a second attempt. Given the special publicity problems surrounding the 1992 election, as well as a number of encouraging signs including the closeness of the vote, the strong editorial support and encouragement of *The Oregonian,* and the undiminished enthusiasm of those who have been working in the program, the chances for passage on a second attempt seem very good. The second attempt is likely to occur in a special election in either the fall of 1993 or the spring of 1994, when other issues will not divert media and voter attention.

Acknowledgments

Support for much of the work related to this paper came from the Audubon Society of Portland, the Meyer Memorial Trust, the Metropolitan Service District (Metro), various local jurisdictions in the metropolitan area, several state agencies, the U.S. Environmental Protection Agency, the U.S. Fish and Wildlife Service, and Portland State University.

Acknowledgment is also given to creative brain-storming sessions with many people, particularly Esther Lev and Lynn Sharp. Portland State University geography students Karla Peters, A. Paul Newman, Maureen Smith, Scott Augustine, Gary Bishop, and Manette Simpson assisted in performing the work on the inventory. Susan Millhauser prepared the location map.

References

Collins, C. 1990. The greening of Portland, can nature survive the Northwest's urban boom?. *The Sunday Oregonian, Northwest Magazine,* 26 August.

Columbia Region Association of Governments. 1971. Proposals to the Portland-Vancouver community for a metropolitan park and open space system. *The Urban Outdoors.*

Curtis, J. T. 1956. The modification of mid-latitude grasslands and forests by man. In *Man's Role in Changing the Face of The Earth,* ed. W. L. Thomas, Jr. Chicago: University of Chicago Press.

Edelman, A. 1991. Personal conversation with one of the founders of the Forty-Mile Loop Land Trust.

Hiss, T. 1990. *The Experience of Place,* New York: Knopf.

Houck, M. C. 1989. Protecting our urban wild lands: Renewing a vision. Address to the City Club of Portland.

———. 1991. Metropolitan wildlife refuge system: A strategy for natural resources planning. In *Symposium Proceedings: Wildlife Conservation in Metropolitan Environments.* Columbia, MD: National Institute for Urban Wildlife.

Houle, M. C. 1988. *One city's wilderness, Portland's Forest Park.* Portland: Oregon Historical Society Press.

Jenkins, R. 1976. Maintenance of natural diversity: Approach and recommendations. In *Transactions, Forty-First North American Wildlife and Natural Resources Conference.* Washington: Wildlife Management Institute.

Johnson, D. M. 1987. Weather and climate of Portland. In *Portland's changing landscape,* ed. L. W. Price. Portland: Portland State University, Department of Geography.

Land Conservation and Development Commission. 1985. Oregon's statewide planning goals. Salem, OR: Land Conservation and Development Commission.

Lev, E., and M. C. Houck. 1988. Planning for urban wildlife in Metropolitan Portland Oregon. *Women in Natural Resources* 9(3): 23–26.

Lev, E., and L. Sharp. 1991. The Portland-Vancouver natural areas inventory: Field surveys and preliminary wildlife data. In *Symposium Proceedings: Wildlife Conservation in Metropolitan Environments.* Columbia, MD: National Institute for Urban Wildlife.

Little, C. E. 1990. *Greenways for America.* Baltimore: The Johns Hopkins University Press.

Mantell, M. A., S. F. Harper, and L. Propst. 1990. *Creating successful communities, a guidebook to growth management strategies.* Washington: Island Press.

Market Trends. 1990. Public Opinion Survey. Portland: Market Trends, Inc.

Metro. 1989. *Metro recreation resource study.* Portland: Metropolitan Service District.

————. 1991a. Sponsor list. Portland: Metropolitan Service District.

————. 1991b. Metropolitan greenspaces. Program brochure and map. Portland: Metropolitan Service District.

————. 1991c. *Goals for Metropolitan Greenspaces, a report on the goal-setting workshops.* Portland: Metropolitan Service District.

————. 1992. *Metropolitan Greenspaces master plan.* Portland: Metropolitan Service District.

Mumford, L. 1939. Regional planning in the Pacific Northwest, a memorandum. Portland: Northwest Regional Council.

Olmsted Brothers. 1903. Appendix. In *Report of the park board, Portland, Oregon.* City of Portland, Park Board.

Poracsky, J. 1991. The Portland-Vancouver natural areas inventory: Photo interpretation and mapping. In *Symposium Proceedings: Wildlife Conservation in Metropolitan Environments.* Columbia, MD: National Institute for Urban Wildlife.

Poracsky, J., L. Sharp, and E. Lev. 1991. *Metro Natural Areas Inventory, phase II - mapping, field surveys, and database creation: Final report.* Portland: Metropolitan Service District.

Poracsky, J., L. Sharp, E. Lev, and M. Scott. 1992. *Metropolitan Greenspaces Program data analysis, part 3: Summary conclusions and recommendations.* Portland: Metropolitan Service District.

Rogers, R. T., and M. C. Houck. 1984. Wetlands: Dirt cheap and disappearing. *Landmark* 1:12–19.

U.S. Bureau of the Census. 1991a. *1990 census of population and housing, summary population and housing characteristics, Oregon.* CPH-1-39. Washington: U.S. Government Printing Office.

————. 1991b. *1990 census of population and housing, summary population and housing characteristics, Washington.* CPH-1-49. Washington: U.S. Government Printing Office.

Vesterby, M. 1987. Land use change experiences from pilot studies near Portland, Oregon. In *Sustaining Agriculture Near Cities,* ed. W. Lockeretz. Ankeny, IA: Soil and Water Conservation Society.

Ecology Education for City Children

Karen S. Hollweg

Introduction

Sustainable cities are based on natural systems. But in the United States today, a variety of actions are needed to preserve and restore our cities' biodiversity. Education is fundamental to this objective. We need, in our democratic society, citizens who understand ecological concepts and make wise decisions based on these concepts. But what, in fact, do our citizens know and value? Are today's city dwellers aware of the diversity of organisms living around them? Do they understand populations and their ecological interactions? Do they understand and appreciate the enterprise of science, the value of inquiry, and the findings that can come from collecting and analyzing data?

The Need for Ecology Education

The media regularly point out the lack of science literacy among Americans. In terms of urban biodiversity, that lack is apparent to me. An overwhelming majority of the children and adults I work with in cities across the country hold the view that there is "nothing alive here." As one Denver elementary school principal told me, "Children think they have to go someplace—like to the mountains or to the site of a TV special—to see nature. They're not aware of what's here in their immediate surroundings."

Even laypeople who have an interest in the natural world focus on individual organisms—the birds that come to their feeders, the "problem species" such as racoons and squirrels, and the "glamour species," like the peregrine falcons released among our office buildings or the deer in our parks and preserves. A 1985 study sponsored by the U.S. Fish and Wildlife Service (Westervelt and Llewellyn 1985) found that the most prevalent attitude toward animals among fifth- and sixth-graders and adults was a "humanistic one," that is, "an emotional identification with the individual animals, mainly pets, combined with strong anthropomorphic tendencies." These people think of individuals, not populations—individuals that are fed and housed by humans, not animals that live in a natural habitat and interact with other organisms in biological communities. That is hardly the viewpoint one would expect people to have who embrace and support, let alone understand, the need for and value of biodiversity and the processes of natural systems.

A useful construct for thinking about the kind of ecologically literate citizenry we need came out of a miniconference in 1988 sponsored by the National Science Foundation (SRI International 1988, 90). As one participant explained: "Scientists, like professional athletes, ultimately need a knowledgeable and appreciative audience to play to, and in sports there is such an audience. Kids grow up playing sports; sports are discussed in the family; they are on television; there are sports sections in the paper. . . . A multitude of informal processes throughout life combine to generate a "sports literate" population. . . . I think this is a useful parallel for thinking about the development of science [and ecological] literacy."

My ideal is to see kids who grow up in cities swinging insect nets to see how many tiny critters they can find, down on their hands and knees observing the behavior of bugs, and standing spellbound watching invertebrate predators nab their prey. And, in neighborhoods lucky enough to have a pond or stream, the kids would strain the water and sift through the muck to discover all the tiny organisms that live there, and to watch food chains function.

They would find that different kinds of plants and animals grow and survive in different habitats. They would talk with each other and their parents about their "finds," and watch experts conducting and learning from similar studies on TV. They would become involved in and supportive of the whole endeavor of learning about and preserving biodiversity in the same way they become involved in and support athletics.

Ways to Meet the Need

How can this be achieved? You might say, the easiest way would be through our schools. Experts continue to advocate hands-on, inquiry-based science for our youngsters—experiences that let them see, touch, smell, describe, ask questions about and investigate the natural world (National Center for Improving Science Education 1989). But textbooks and lectures increasingly dominate the teaching of science at all grade levels. National surveys show that the amount of time students spend experiencing hands-on science in our schools is *decreasing* (Weiss 1987).

We have learned that reversing this trend is no simple matter. In the post-Sputnik era of the sixties and early seventies, we spent $100 million in an attempt to upgrade science education. We developed new science curricula that emphasized hands-on experiences and retrained teachers to use the new curricula. We got about 25 percent of our schools to use the new curricula (Mechling and Oliver 1983). Today's estimates are that no more than 5 percent of our teachers are teaching hands-on, inquiry-based science (McCormick 1989, p. 3).

The demands on today's classroom teachers are onerous. The logistical complications of amassing enough materials for a class of twenty-five to thirty children and supervising that many youngsters outdoors as they excitedly run around with nets or strainers are tremendous. Teachers cannot be expected to provide their students with hands-on outdoor science experiences on their own.

The Denver Audubon Project

Denver Audubon's Urban Education Project is one answer to this dilemma. In the early 1980s Denver Audubon Society's board agreed that everyone should have opportunities to get outside, search for the creatures that live there, and experience the thrill of finding something special. We felt that city kids need to be encouraged to feel the textures of leaves, to explore their natural environment, and to find out what spiders really look like. They need chances to net fast-moving insects and discover how they live. We believed that children who feel the fascination of their environment are likely to want to see it cared for. But our needs assessment data indicated that most children growing up in Denver lacked these kinds of experiences. They were not aware of their surroundings.

To address this need, Denver Audubon started its "Urban Education Project." It provides school-age children with opportunities to explore the plants and animals in their own neighborhoods. Trained volunteers teach the children where to find things and how to observe them. Volunteers choose from a variety of prepared *Outdoor Biology Instructional Strategies* (OBIS) (Lawrence Hall of Science, University of California, 1981), and conduct hour-long sessions with small groups. Each volunteer works with about six youngsters. Most of their time is spent searching for, observing, and discovering new things. Kids explore the natural world in city parks, schoolyards, rights-of-way, and vacant lots. They find different kinds of roots and creatures that live under the soil, build bird feeders to attract and observe urban wildlife, and do experiments to test the food preferences of neighborhood birds. They investigate the interrelationships between plants and animals, and have chances to compare their findings with one another. For some children this experience may lead to further study; for others, to a recreational interest. All participants gain a greater awareness of their surroundings, a better understanding of the natural world and our relation to it.

We started in 1985 with 250 eight- to twelve-year olds and 60 volunteers, and grew to the point where we now consistently reach about 1,500 children a year at schools and community centers throughout metropolitan Denver. About 45 percent of the children represent racial minorities (African-American and Hispanic) and about 40 percent are from families living at or below poverty level. Each child participates in an average of five explorations.

Approximately 180 volunteers, whose efforts are coordinated by one full-time staff member, make all this happen. Some volunteers spend evenings phoning other Audubon members and friends to raise money and recruit more volunteers. Others help solicit funds from their employers and local corporations. Some volunteers come to the office to bring in materials or help organize boxes of equipment. Others sew or put handles on insect nets at their homes. Even the volunteers who work with kids are involved behind the scenes. They participate in a three-hour training workshop at the beginning of each season to become familiar with the activities and learn new teaching strategies.

Denver's Urban Education Project has an annual budget of about $30,000, all raised from local sources. A publication entitled *Volunteers Teaching Children: A Guide for Establishing Ecology Education Outreach Programs* (Hollweg 1991) describes how the project was started, what is done, who is involved, and how the whole thing is organized. (This book is available from NAAEE, Box 400, Troy, Ohio 45373.)

Similar Projects in Other Cities

Now other cities have started similar programs (see fig. 1). In 1988–91, with the support of a $380,000 National Science Foundation Dissemination grant, we were able to help seven new cities get started and learn from Denver's experiences. The grant paid for each program's one-time start-up costs, but all on-going costs are paid with local funds. Additional funding in 1992–94 supported further expansion.

It has been exciting to see the diverse ways in which these programs have developed. Each program is designed to utilize available local resources, and therefore started with volunteers from different sectors. In Prescott, Arizona, the volunteers largely came from the town's numerous active retired professionals and from the parents of fourth-grade pupils. Seattle's volunteers were a combination of parents, Audubon members, and college students. Louisville's program was built on the concept of "kids teaching kids." Trained high school students went to feeder elementary schools to conduct the investigations. By 1992, every city's program used at least some high-school teens to lead the small groups of children.

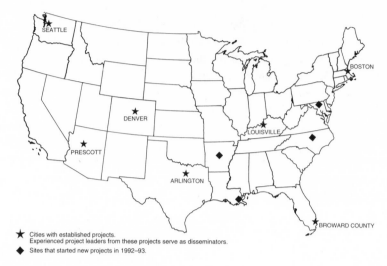

★ Cities with established projects.
  Experienced project leaders from these projects serve as disseminators.
◆ Sites that started new projects in 1992–93.

Figure 1   Urban Education Project Sites

The program sponsors or lead agencies vary from Audubon chapters to the Museum of History and Science in Louisville, Boys and Girls Clubs in Texas, and nature and science centers in Durham and Baltimore.

One of these cities has volunteers coordinating their program. But it seems that a paid coordinator (at least half-time) is the most effective way to run such a program, assuming, of course, that local funds can be raised to pay that person's salary.

As would be expected, the invertebrate resources of Fort Lauderdale (Florida) and New Orleans (Louisiana) are much richer than those of Denver or Boston, especially in January and February. But each city has identified a collection of OBIS investigations that they can use.

Project Outcomes

As children find out how many different species of plants and animals live in a grassy area, discover that different kinds of organisms live in dry and in moist leaf litter, learn that different biological communities thrive on juniper bushes than on lilac bushes, and watch spiders and predatory insects catch their prey, they discover that, as the kids put it, "There are all sorts of things living here that I never noticed before." They begin to gain a concept of the large numbers of organisms, of populations and the different niches they occupy, and to understand why so many more different forms of life are found in a vacant lot or a natural area than in a stretch of chemically treated turf that is a monoculture. They also have a chance to experience the thrill of discovery and to understand and appreciate the work of biologists and ecologists. And just as important, they enjoy sharing their "finds" with others in their group and develop special relationships as they work together.

An incident that occurred with a participating class of fifth-graders in Texas illustrates the value of these experiences. The students had been showing me the strainers, scoops, plankton nets, and other equipment they had made and used to explore a nearby pond, and had been telling me about all the different kinds of tiny aquatic plants and animals they had discovered but previously had no idea were there. Playing devil's advocate, I said, "But why should you be spending class time on this sort of stuff?" Their answers:

> [This program] teaches you what you don't learn in ordinary classrooms. It's more educational than just sitting in a classroom reading out of a book. If you're reading out of a book, with lots of kids it goes in one ear and out the other—but when you're actually doing it, it's more fun, and the kids understand more . . . Like, seeing is believing! . . . And besides it makes you understand the news, like the Exxon Valdez oil spill. On TV they show the otters and birds covered with oil. But there's got to be lots of plankton there too. And now we know that all that little stuff is out there and it's being affected too.

Those children have had experiences with the natural world in their home place that have given them an understanding of biological communities and ecological concepts.

The Future

Nationwide in 1991–92, programs in eight cities reached almost seventy-five hundred city children (about 40 percent disadvantaged minorities), but that is only a beginning. We need many more efforts like these. The cities now sponsoring urban ecology education programs are linked together through the VINE (Volunteer-led Investigations of Neighborhood Ecology) Network. The network enables program coordinators across the country to exchange ideas and collaborate in joint efforts. We have produced a video tape that describes our projects and a portfolio of evaluation instruments and techniques that any city can use to assess its program and student outcomes. We are anxious to reach out and involve others, and think the climate is right for expanding. Those interested in starting a similar program in their city may obtain further information from the VINE Network, c/o the North American Association for Environmental Education (NAAEE), Box 400, Troy, Ohio 45373.

Our society is awakening to the need to improve our children's learning opportunities and to involve parents and community members in education. There are many hopeful signs. I have been especially encouraged by the efforts of the Ecological Society of America's Education Committee. The ecologists, as well as members of other professional organizations such as the American Association for the Advancement of Science (AAAS), are becoming increasingly active in public schools to improve science education. We must persevere and engage more people in our efforts.

Acknowledgment

This article is based upon work supported by the National Science Foundation under Grants No. MDR-8850604 and MDR-9155423.

References

Hollweg, K. S. 1991. *Volunteers teaching children: A guide for establishing ecology education outreach programs.* Denver: Denver Audubon Society.

Lawrence Hall of Science, University of California. 1981. *Outdoor biology instructional strategies (OBIS).* Nashua, NH: Delta Education.

McCormick, K. 1989. Battling scientific illiteracy: Educators seek consensus, action on needed reforms. *Association for Supervision and Curriculum Development Curriculum Update.* June:1–7.

Mechling, K. R., and L. Oliver. 1983. Activities, not textbooks: What research says about science programs. *Principal* 62:40–43.

National Center for Improving Science Education. 1989. *Getting started in science: A blueprint for elementary school science education.* Washington: The Center.

SRI International. 1988. *An approach to assessing initiatives in science education: Volume 2, pilot assessment of the National Science Foundation's investments in informal science education.* Menlo Park, CA: SRI International.

Weiss, I. R. 1987. *Report of the 1985–86 national survey of science and mathematics education.* Research Triangle Park, NC: Research Triangle Institute.

Westervelt, M. O., and L. G. Llewellyn. 1985. *Youth and wildlife.* Washington: U.S. Fish and Wildlife Service.

Conclusion

Crosscutting Themes and Recommendations

Rutherford H. Platt

The sixteen contributions to this volume offer, compactly, a diversity of experience and insight regarding the values and means of preserving biodiversity in North American cities. As in the symposium from which this work emerged, the primary concern of the authors has been the maintenance and restoration of aquatic and terrestrial biotic systems within cities or newly developing areas. This collection of papers has sampled the rich array of scientific, legal, and geographical issues encompassed under the rubric of urban biodiversity.

In and among the diverse perspectives and case histories related in the foregoing articles, several crosscutting themes may be identified that help to integrate this seemingly disparate collection. Such themes, which appear to be most fruitful, as discussed below, include the following:

> Changing perceptions of nature in cities;
> Multiple uses of natural spaces in cities;
> Alternative scales of intervention;
> Institutional flexibility and innovation;
> The role of science in urban design.

Changing Perceptions of Nature in Cities

In my introduction to this volume, I quoted from *The Greening of the Cities* by David Nicholson-Lord (1987, 82): "The picture that emerges is thus one of discovery, of an urban society beginning to look at its immediate surroundings with fresh eyes, seeing new possibilities in old things. A radical change in perception is involved."

Several articles in this volume have reflected such a "radical change of perception" regarding the commonplace and familiar. Landscape architect Michael Hough, drawing on themes from his book *Out of Place: Restoring Identity to the Regional Landscape*, refers to the discovery that vacant lots, railroad rights-of-way, cemeteries, and other open spaces scattered around cities comprise "fortuitous landscapes." If left to their own devices, such incidental bits of unbuilt land may nurture remnants of indigenous flora and fauna, innocuously blended with exotics in some cases.

Hough's perspective is shared by fellow landscape architects Jack Ahern and Jestena Boughton, who write that "paradoxically, the landscape industry has become a major obstacle to the acceptance of a new landscape aesthetic founded explicitly on sustainable native species and natural processes." They urge the replacement of manicured, nonsustainable landscaping where possible with patches of low-maintenance wildflower meadows (as in highway median strips). Changing perceptions also influence the design, management, and use of public spaces, as documented in Annaliese Bischoff's case study of Forest Park in Springfield, Massachusetts, over the past century.

Aquatic ecologists Marjorie Holland and Raymond Prach note the importance of ecotones, or boundaries, between discrete ecological communities as zones for the transfer of nutrients, sediments, and energy, processes of particular interest in landscape design and management. Orie L. Loucks, also a wetlands ecologist, refers to the need to enhance the "regenerative capacity" of ecological systems in urban areas, a capacity that presumably depends in part on the recognition and protection of ecotones.

The relationship of land use in drainage basins to water quality in recipient water bodies is well recognized today, but for long that has not been the case. Loucks's research at Lake Wingra in Wisconsin in the early 1970s documented the comparative effects of urban and rural land-surface runoff on water quality. On a larger scale, Charles Goldman has studied the effects of changes in land use in the Lake Tahoe basin on the clarity and productivity of that lake. These types of studies have helped to broaden the geographical perception of water quality and wetland management to include urbanizing uplands that drain into them.

Another perceptual breakthrough has occurred with the recognition of the need for connectivity between patches of natural habitat—isolated plots of ground are insufficient for many plants and animals to survive. In their research at the incredibly biodiverse Indiana Dunes National Lakeshore, Richard Whitman and his colleagues have found that "the trend toward isolation and fragmentation [due to urbanization in the vicinity] has numerous and often compounding effects on the viability of the protected areas as a system and on their biological value. For instance, the reduced ingress of individual animals of plant propagules into the protected area jeopardizes population viability, the maintenance of genetic diversity, and the likelihood of recolonizing an area following some type of disturbance."

Similarly, Timothy Beatley reports that efforts to protect the Coachella Valley fringe-toed lizard through a regional habitat conservation plan confronted a dilemma as to what amount and spatial distribution of desert habitat would be sufficient to ensure the lizard's survival. The natural areas inventory prepared for the Metropolitan Service District in Portland (Oregon), as discussed by Joseph Poracsky and Michael Houck, also reflects the importance of linkage between patches of habitat.

Other writers remind us that perceptions of nature in the city may arise at a very personal, nonscientific level. John Dwyer and his colleagues report their findings on the "deep emotional ties between people and trees" that defy easy classification. They cite research by Roger Ulrich (1984) on the therapeutic value of scenes of vegetation to hospital patients. They also quote Donald Appleyard (1980): "Trees comprise one of the last representatives of nature in the city, providing a constant reminder of the natural world beyond as well as our distant past." Karen Hollweg in her paper suggests that the most humble plots of unpaved ground or even mud puddles may provide excitement for children and adults who learn to look carefully at the organisms and "critters" found therein.

Multiple Uses of Natural Spaces in Cities

Unlike ecological preservation in remote locations, urban biodiversity must coexist with and often depend upon anthropocentric activities. It is axiomatic today that unbuilt land within urban areas must serve multiple functions. Frederick Law Olmsted, Sr.'s, "Emerald Necklace" plan for the Boston park system of the 1880s, which provided recreation, amenity, and hydrologic benefits, was the prototype of multifunction designs for open spaces. In a literal extension of an Olmsted plan (the Forty-Mile Loop proposed for Portland, Oregon, by John Charles Olmsted in 1903), the Portland Metro natural areas inventory described by Joseph Poracsky and Michael Houck would expand the existing system to encompass new areas and ecological functions. Timothy Beatley notes that the program to save habitat for the fringe-toed lizard incidentally has conserved areas of open desert that will provide recreational and scenic benefits, and possibly mitigate flash flood hazards, for the quickly urbanizing Coachella Valley.

Urban greenways oriented to stream valleys, shorelines, upland ridge lines, or other linear physical features, as documented elsewhere by Charles E. Little (1990), offer a variety of opportunities for achieving multiple benefits. Ann Riley's account of the tortuous evolution of a multifunction plan for the Wildcat and San Pablo creeks in the East Bay area of California suggests how the various stakeholders must find collaborative solutions for the management of urban stream corridors. In that case, flood and erosion control brought additional benefits in biodiversity, recreation, and scenic amenity through the replacement of "trapezoidal" channel engineering with a more ecologically sensitive design.

Urban wetlands may similarly serve multiple purposes, as in the case of Wascana Centre in Regina, Saskatchewan. According to Holland and Prach, that 2,300-acre lake-wetland complex is managed for recreation, environmental education, and waterfowl habitat, and also to provide a source of nonpotable irrigation water for municipal needs. Separate ponds of the lake system are managed differently, one for contemplative use and the other for active recreation depending upon the season.

Annaliese Bischoff's Forest Park case study indicates that various functions may arise and recede over time according to changing social needs and preferences, and that earlier functions (e.g., food production and forest management) may be worth reviving as elements of urban sustainability.

Alternative Scales of Intervention

On a theme closely related to the preceding discussion, the experiences documented in this volume reflect the broad range of geographic scales at which management efforts may seek to "protect and restore urban biodiversity." Dwyer's interest in individual trees and Loucks's example of the dragonfly pond in Yokahama, Japan, bound the spectrum at the micro end. The Lake Tahoe basin regional planning program (Goldman), the Coachella Valley habitat conservation plan (Beatley), the Portland Metro natural areas inventory (Poracsky and Houck), and the Indiana Dunes National Lakeshore (Whitman et al.) represent the macroscale of urban resource management. The mesoscale is represented by the Wascana Centre complex (Holland and Prach), the Wildcat-San Pablo creeks system (Riley), and the DesPlaines River wetland restoration project (Hey).

Generalization regarding the optimal scale of intervention is hazardous. One would expect greater political and scientific complexity with broader scale, but perhaps also greater-scale economies and a wider range of benefits. Much depends upon who is in a position to intervene, legally and geographically, and what the primary goals for intervention are. From the standpoint of public education, small-scale models of ecological restoration such as the dragonfly pond may be very cost-effective and inspire imitation through nongovernmental efforts. But most of the experiences related above involve larger-scale project areas that do not fit conveniently within the geographic territory of a single resource-management entity. If efforts to protect and restore urban biodiversity are to be driven by ecological reality rather than practicality, the crucial need is for management institutions to adapt to science, rather than vice versa.

Institutional Flexibility and Innovation

The American federal system of government is poorly adapted to the management of urban ecological resources. Political authority is divided among the three primary levels of government: federal, state, and county/municipal. There is little or no correspondence between the territories of these political divisions and the spatiality of ecological phenomena, nor for that matter do either of these correspond to the geographic extent of most metropolitan urban regions. Furthermore, even within a particular level of government, e.g., the state, management authority is fragmented by function among competing administrative agencies such as natural resources departments and economic development agencies.

Adapting political institutions to science therefore involves, in most instances, (1) a coalition of existing management units, or (2) the establishment of a new public or quasi-public unit of authority, or (3) both. Instances of shaping government institutions to conform with natural phenomena have been widely documented by Charles H. W. Foster (1986), under the rubric of bioregionalism. The papers in this collection relate many examples of institutional flexibility. Indeed, among those authors who address institutional issues at all, virtually all of them deal with innovative approaches, in contrast to reliance upon the preexisting units of government. (It will be admitted, however, that the papers were not invited randomly, and that the lead editor's interest in institutional management of resources was a major criterion for selection.)

The two largest-scale institutions considered in this volume, the Indiana Dunes National Lakeshore (Whitman et al.) and the Tahoe Regional Planning Agency (Goldman), reflect two different approaches to the protection and restoration of biological habitats in urbanizing regions. The former involves federal ownership and management of critical habitats by a unit of the National Park System specially created, after decades of controversy, by Congress. The authorized boundaries of the National Lakeshore in fact were drawn to include most of the remaining patches of natural dunes habitat at the time of its establishment in the mid-1960s (Platt 1972; Engel 1983). Management of the Lake Tahoe basin, by contrast, has been addressed through a bistate regional planning agency established by an interstate compact between California and Nevada. The agency has limited authority to influence land development and impacts on water quality in the basin, but at least it speaks with a regional voice for a resource that is artificially bisected by a state boundary.

The Portland Metropolitan Service District, discussed by Poracsky and Houck, exemplifies the adaptation of an existing regional special district to encompass new functions relating to biodiversity. The Portland Metro District is one of the few multipurpose urban regional districts in the United States. According to a recent article in *The Atlantic*, "the Metropolitan Service District or 'Metro' brings urban and suburban interests together in a unique popularly elected government covering parts of three counties" (Langdon 1992). Apart from such operational responsibilities as the airport and convention center, Portland Metro contributes widely to the habitability of its region though its greenspaces program and its recent sponsorship of the regional natural areas inventory. Finding that even its own regional boundaries were insufficient to encompass the relevant natural systems, Metro has collaborated with additional counties, both in Oregon and across the Columbia River in the state of Washington, in conducting the inventory and proposing new areas to be protected.

In Canada, the Metropolitan Toronto and Region Conservation Authority, mentioned by Hough, plays a roughly equivalent role as a broadly conceived regional resource-management agency. While it has not undertaken an inventory as ambitious as Portland Metro's, the Metro Toronto authority owns and manages a number of stream valleys that comprise the region's primary system of multiuse natural areas. At a more local scale, the Wascana Centre Authority (Holland and Prach) was established in 1962 by the province of Saskatchewan with an eleven-member management committee composed of representatives of the city of Regina, the province, and the University of Regina.

For the more common situation where no inherently flexible regional institution exists, new arrangements must be developed on an ad hoc basis. But "ad hockery" does not simply happen spontaneously. One or more public or quasi-public prime movers must take the initiative in forging multiparty approaches. In Beatley's case study of the Coachella Valley, the U.S. Fish and Wildlife Service catalyzed the formation of the "lizard club," which included a number of federal, regional, nonprofit, and native American participants. Conversely, coalitions of unlikely bedfellows may arise in opposition to a proposal of a conventional resource management entity. Thus, according to Riley, the Army Corps of Engineers' plan to line Wildcat and San Pablo creeks with concrete provoked years of negotiation involving over a dozen public and private entities and eventually yielding a "greener" solution.

The role of the individual in stimulating institutional innovation must not be overlooked. The key participant in the Des Plaines River wetlands restoration project described by Donald Hey is Donald Hey himself. Through his nonprofit Wetlands Research, Inc., he has personally catalyzed a unique intergovernmental approach to habitat restoration. While the legendary figures of conservation history—e.g., John Muir, Aldo Leopold, Robert Marshall, and David Brower—were largely concerned with remote wilderness areas, the struggles for urban biodiversity require no less in personal commitment and initiative and much more in grassroots political savvy and diplomacy.

The Role of Science in Urban Design

Throughout this book, the writers have directly or indirectly referred to the need to better apply the insights and methods of the natural sciences to the planning of urban regions, communities, and neighborhoods. Loucks in his conclusion mentions "the substantial scientific underpinnings required to develop and maintain regenerative capacity in cities. When we intervene to try to restore sustainability, we need to be sure the outcome will be an improvement, not a higher risk of impoverishment."

A primary role of science in the management of urban biodiversity is to delineate boundaries, or ecotones, between discrete natural subsystems, as argued by Holland and Prach. This is not an easy task. The administration of the federal wetlands program (Section 404 of the Clean Water Act) has been greatly impeded by ambiguous wetland indicators, according to James Schmid. The Portland natural areas inventory was greatly facilitated by the use of geographic information systems that plot the spatial incidence of multiple scientific and management variables (Poracsky and Houck). One of the chief obstacles to preparing a plan to save the Coachella Valley fringe-toed lizard was uncertainty as to the lizard's territorial range and its ability to adjust to encroachments on its habitat (Beatley). Similar issues were reported by Whitman and his colleagues regarding flora and fauna in the Indiana Dunes National Lakeshore.

Even where delineation of the management unit is not at issue, as with the Lake Tahoe basin, the prediction of off-site and cumulative impacts of human activities within that territory poses a major, ongoing scientific challenge, according to Goldman. The predictability of adverse (or favorable) ecological effects of changes in land use is a problem cited by several authors.

Not only must the development process be managed to conserve ecological phenomena, but those phenomena themselves frequently require nurturing through human intervention guided by the advice of scientists. Ahern and Boughton refer to the need for selective and properly timed mowing to prevent wildflower meadows from reverting to scrub woodlands. Remnant prairies within the Indiana Dunes National Lakeshore, according to Richard Whitman and his colleagues, are periodically burned under carefully controlled conditions to emulate the effects of natural fires. Hey's wetlands restoration project involves minute regulation of water flow to maintain an appropriate balance of nutrients in the recovering wetland. Gregory McPherson outlines fairly precise ways to design tree planting for urban streets to maximize microclimate benefits. Ann Riley mentions new methods for stabilizing eroding stream banks through the well-informed selection and placement of riparian vegetation.

Of course, the solicitation and application of scientific advice in the design and redesign of urban communities return us to the question of institutional responsibility. Unless there is a manager or coalition of managers who ask the right questions and put the best available answers into practice, the fruits of scientific research will wither on the vine.

Environmental education is prerequisite to the ability of society to establish appropriate institutions for asking scientists the right questions at the right time. As the landscape architect Barrie M. Greenbie (1990) has put it, "City people must be able to experience nature continuously in the circadian rhythms and spatial webs of their daily lives. A child can learn at least as much, if not more, about nature from a backyard garden or . . . bird feeder as from a nature awareness center."

The programs described by Karen Hollweg to acquaint schoolchildren with the bugs and "critters" in their own neighborhoods and backyards are essential to this process. The Chicago schoolchildren exposed to Walter Moody's text on city planning grew into the voters who approved the bond issues to realize much of Burnham and Bennett's 1909 *Plan of Chicago*. Today's children who discover urban biodiversity will be the voters, scientists, politicians, and activists of the next century.

Future Directions

The crosscutting themes identified above provide a guide to the types of recommendations for future action that logically flow from this set of papers. In summary form, these include the following:

(1.) A national blue ribbon committee, comparable to the Outdoor Recreation Resources Review Commission established by Congress in 1958, should be created to review the functions and benefits of biodiversity in urban areas and to assess the status of efforts to achieve it.

(2.) The federal Land and Water Conservation Fund Act of 1965, which provides matching grants for open space acquisition and development, should be updated and reauthorized to emphasize projects within metropolitan statistical areas (MSAs) in preference to nonmetropolitan sites.

(3.) In the tradition of Patrick Geddes's "open spaces study" of Edinburgh, an effort should be made to *identify and inventory existing open spaces* regardless of ownership within urban areas. A model for the development of a computerized data base on natural areas is provided by the Portland, Oregon, Metro inventory described by Poracsky and Houck.

(4.) For spaces in public ownership (e.g., surplus lands, watershed lands, tax-delinquent properties, etc.), the responsible governmental unit should develop a "minimal management" strategy. Community or "allotment" gardens should be encouraged for agricultural or horticultural purposes where feasible. Other tracts should be set aside for the rejuvenation of native biotic species and communities. Informal public access should be encouraged where it is consistent with considerations of personal safety.

(5.) Tracts of public open space may be "adopted" by community and civic organizations, local conservation groups ("Friends of ———"), churches, school classes, and similar groups for purposes of maintenance, clean-up, interpretative programs, and other activities. Local organizations often assume responsibility for the clean-up of segments of highways. They could undertake projects to promote biodiversity as well as pick up trash.

(6.) Owners of undeveloped parcels should be assisted to understand the natural phenomena and processes found on their land. Restoration of biodiversity on degraded private tracts should be encouraged through an urban equivalent of the U.S. Department of Agriculture Cooperative Extension Program. Where public access to private land is desirable, such parcels should be publicly acquired in fee or easement from the owner.

(7.) Public awareness of watersheds and drainage systems should be enhanced through better signage identifying streams, wetlands, floodplains, and divides between major drainage basins. Explanatory maps and educational displays on local hydrology and biodiversity should be located at suitable sites along urban highways.

(8.) Finally, interaction between urbanists and natural scientists as reflected in this book, should become the norm rather than the exception as we collectively seek to respond to the challenges of living in a world whose population is more than half urban.

References

Appleyard, D. 1980. Urban trees, urban forests: What do they mean? In *Proceedings of the National Urban Forestry Conference*, 13–16 November 1978. Syracuse: State University of New York College of Environmental Science and Forestry.

Engel, J. R. 1983. *Sacred sands: The struggle for community in the Indiana Dunes.* Middletown, CT: Wesleyan University Press.

Foster, C. H. W. 1986. Bioregionalism. *Renewable Resources Journal* 4(3) (Summer): 12–14.

Greenbie, B. M. 1990. Synthesizing the oxymoron: The city as a human habitat. Unpublished discussion paper presented at the Sustainable Cities Symposium, Chicago Academy of Sciences.

Hough, M. 1990. *Out of place: Restoring identity to the regional landscape.* New Haven: Yale University Press.

Langdon, P. 1992. How Portland does it. *The Atlantic* 270(5): 134–42.

Little, C. E. 1990. *Greenways for America*. Baltimore: The Johns Hopkins University Press.

Nicholson-Lord, D. 1987. *The Greening of the cities*. London: Routledge & Kegan Paul.

Platt, R. H. 1972. *The open space decision process*. Research paper no. 142. Chicago: University of Chicago Department of Geography.

Ulrich, R. S. 1984. View through a window may influence recovery from surgery. *Science* 224:420–21.

Notes on the Editors and Authors

Jack Ahern is an associate professor in the Department of Landscape Architecture and Regional Planning at the University of Massachusetts, Amherst. He has a master's degree in landscape architecture from the University of Pennsylvania, where he was introduced to ecological planning and design by Ian McHarg. He has spent much of his professional life involved with the application of ecological principles to landscape planning and site scale design.

Timothy Beatley is an associate professor in the Department of Urban and Environmental Planning in the School of Architecture at the University of Virginia. He heads the environmental planning curriculum there, teaching environmental planning and policy, environmental ethics, biodiversity conservation, and coastal management. Beatley holds a Ph.D. in city and regional planning from the University of North Carolina at Chapel Hill.

Annaliese Bischoff, formerly Ann Marston, is an associate professor of landscape architecture in the University of Massachusetts Department of Landscape Architecture and Regional Planning. She also serves as associate director for the University Writing Program and the Honors Program. She recently spent six months in Berlin under a Fulbright Research Fellowship, where she conducted research on the design of public spaces and historic sites. She received an M.L.A. from the College of Environmental Science and Forestry at the State University of New York in Syracuse, and a B.A. from Brown University.

Jestena Boughton is an assistant professor of landscape architecture at the University of Massachusetts, Amherst. She earned her masters' in landscape architecture at the University of Pennsylvania. For several years she worked in the Seattle landscape architecture and planning firm of Jones & Jones. A wildflower meadow she designed on 100 acres reclaimed from an airfield on Lake Washington continues ten years later to provide shorebird habitats and a low-maintenance setting for the National Oceanic and Atmospheric Association (NOAA) Western Regional Center. She also directed the design and construction of the Tashkent Peace Park in the former USSR. She is president-elect (1992) of the Boston Society of Landscape Architects.

Kenneth L. Cole received his doctorate in geology from the University of Arizona. He was the plant research ecologist at Indiana Dunes National Lakeshore from 1984 until 1991. He is presently working as global change coordinator for the National Park Service's unit at the University of Minnesota. His research interests include paleoecology, fire ecology, and palynology.

John F. Dwyer is the project leader for recreation research of the North Central Forest Experiment Station, Chicago, Illinois. His research includes policy issues and problems concerning the management and use of forest resources, with a specific focus on the recreational use of forests near populated areas. Dwyer serves or has served as associate editor of the *Journal of Leisure Research, Journal of Leisure Studies,* and *Journal of Arboriculture.*

Daniel B. Fagre received his doctorate in animal ecology from the University of California at Davis. He was the animal research ecologist at the Indiana Dunes National Lakeshore from 1988 until 1991. His research interests are in landscape ecology and its application to various attributes of animal communities. He is presently working as global change coordinator at Glacier National Park.

Paul H. Gobster is a research social scientist with the USDA Forest Service's North Central Forest Experiment Station in Chicago. Gobster holds degrees in recreation planning, landscape architecture, and environment-behavior studies from the University of Wisconsin. Before joining the Forest Service in 1987, he worked as a natural resource planner and a professor of landscape architecture. His current research interests include urban nature access and the aesthetic experience of sustainable ecosystems.

Charles R. Goldman is a professor of limnology and the former chairman of the Division of Environmental Studies at the University of California at Davis. Goldman has received many awards including the National Science Foundation Senior Postdoctoral Fellowship in 1964 for limnological research in the Arctic and a Guggenheim Fellowship in northern Italy. The "Goldman Glacier" in Antarctica was named in his honor in 1967. Goldman's single most important and sustained contribution is his thirty-three years of research on Lake Tahoe. Dr. Goldman is the director of the Lake Tahoe Research Group and was recently senior scientist for a National Geographic expedition to Lake Baikal. He has written four books and 360 scientific publications, and has produced four documentary films which are in worldwide distribution.

Donald L. Hey received his B.S. degree in civil engineering from the University of Missouri, an M.S. in water resources engineering from Kansas University, and a Ph.D. in environmental engineering from Northwestern University. He serves on the National Research Council's Committee on the Restoration of Aquatic Systems. He is the director of Wetlands Research, Inc. the organization responsible for the Des Plaines River Wetlands Demonstration Project.

Marjorie M. Holland is currently an associate professor of biology at George Mason University in Fairfax, Virginia. She is the director of the Public Affairs Office for the Ecological Society of America. She also has consulted for the Man and Biosphere Programme at UNESCO, was an assistant professor of biology at the College of New Rochelle, and was the executive director of the Water Supply Citizens Advisory Committee with the Metropolitan District Commission in Boston, Massachusetts.

Karen S. Hollweg is currently the director of the VINE (Volunteer-led Investigations of Neighborhood Ecology) Network, which encourages cities in carrying out education programs in urban ecology and supports the start-up of such programs in new cities. In 1984, she started Denver Audubon Society's Urban Education Project, which has involved approximately 1,500 inner-city youth and 180 volunteers annually. She is the coauthor of *Investigating Your Environment,* which she field-tested while working at Biological Sciences Curriculum Study (BSCS). Hollweg is also the coauthor of *A Study of the Ecology of the Prairie: A Unit for the Seventh Grade Life Science Course,* and *Primary Integrated Curriculum: Grade One.*

Michael C. Houck received his B.S. degree in zoology from Iowa State University in 1969 and his M.S.T. in biology from Portland State University (Oregon) in 1972. During the 1970s, he directed the Community Research Center at the Oregon Museum of Science and Industry, and taught high school biology at Oregon Episcopal School in Portland. Since 1980, Houck has pursued urban wildlife habitat issues through the Audubon Society of Portland, where he is the urban naturalist for the society's Metropolitan Wildlife Refuge System project. Houck is the director of the Urban Streams Council, a program of The Wetlands Conservancy, a private, nonprofit land trust.

Michael Hough is a partner in the Toronto landscape architectural firm of Hough, Stansbury and Woodland Limited, which he founded in 1964. He has acted as chief researcher and designer for a number of urban reforestation and naturalization projects for the National Capital Commission in Ottawa, has represented his firm in the team that produced the Toronto Harbourfront 2000 plan in 1987, and is working with the Waterfront Regeneration Trust in Toronto. Hough is also a professor of environmental studies at York University in Toronto. Recent books include *City Form and Natural Process* and *Out of Place: Restoring Identity to the Regional Landscape,* both published by Yale University Press.

Orie L. Loucks is Ohio Eminent Scholar in Applied Ecosystem Studies and a professor of zoology at Miami University, Oxford, Ohio. His training includes B.Sc.F. and M.Sc.F. degrees in forestry from the University of Toronto and a Ph.D. in botany from the University of Wisconsin-Madison. He joined the Department of Botany at the University of Wisconsin in 1962. From 1969 to 1973, he headed an interdisciplinary study of the largely urban Lake Wingra watershed, part of the U.S. contribution to the International Biological Program. In 1978, he joined The Institute of Ecology (TIE) in Indianapolis as science director and headed a series of studies on the regional effects of air pollutants and acidic deposition. In 1983, Loucks became the director of the Holcomb Research Institute at Butler University in Indianapolis. In the mid-1980s, he was a member of the National Academy of Sciences (NAS) Board on Water Science and Technology, and was U.S. co-chair of the joint National Research Council-NAS/Royal Society of Canada study reviewing the 1978 Great Lakes Water Quality Agreement. His most recent relevant publications include "Looking for Surprise in Managing Stressed Ecosystems" in *Bioscience,* and *Natural Diversity as a Scientific Concept in Resource Management* (Proceedings of the Workshop).

E. Gregory McPherson is a research forester with the U.S. Department of Agriculture Forest Service in Chicago. Prior to holding this position, he was an associate professor of landscape architecture in the School of Renewable Natural Resources at the University of Arizona in Tucson. He received a bachelor's degree in general studies at the University of Michigan, a master's in landscape architecture at Utah State University, and a Ph.D. in urban forestry at the State University of New York College of Environmental Science and Forestry in Syracuse. He edited the book *Energy Conserving Site Design,* which won a Merit Award in 1985 from the American Society of Landscape Architects, and is the coauthor with Dr. Charles Sacamano of a recent book entitled *Southwestern Landscaping that Saves Energy and Water.* His research is aimed at quantifying the effects of urban vegetation on the physical environment of cities.

Pamela C. Muick is a Science and Diplomacy fellow with the American Association of Science in the Asia Bureau of the U.S. Agency for International Development, Washington, D.C. She is coauthor of a recent award-winning book entitled *Oaks of California.* Her research interests include forest ecology and restoration. She received her M.S. and Ph.D. from the University of California, Berkeley.

Noel B. Pavlovic is the research plant ecologist at the Indiana Dunes National Lakeshore. He received his masters' degree in ecology from the University of Tennessee and is currently a candidate for the Ph.D. at the University of Illinois at Chicago. His research focuses on rare plant conservation and restoration, and vegetation dynamics.

Rutherford H. Platt is a professor of geography and adjunct professor of regional planning at the University of Massachusetts at Amherst. He holds a B.A. in political science from Yale and a law degree and Ph.D. in geography from the University of Chicago. He is a member of the Illinois Bar. His research, teaching, and consulting have dealt with public policy issues of land and water resources, particularly the management of floodplains, wetlands, and coastal zones. He has served on several national panels including five committees of the National Academy of Sciences/National Research Council. He has written numerous articles and monographs and has recently published a textbook: *Land Use Control: Geography, Law, and Public Policy* (Prentice-Hall, 1991).

Joseph Poracsky is an associate professor of geography and the director of the Cartographic Center at Portland State University, where his primary teaching responsibilities are in the areas of cartography and remote sensing. His Ph.D. is from the University of Kansas, where he also spent seven years as the senior remote sensing application specialist with the Kansas Applied Remote Sensing (KARS) Program. He also served as the cartographic editor for the *Oregon Environmental Atlas,* published in 1988 by the Oregon Department of Environmental Quality. During the past several years, Poracsky has become increasingly involved in work concerning the interaction of urbanization and the natural environment.

Raymond W. Prach is a consultant in environmental planning, information transfer, environmental assessments, and ecological studies. He has a B.Sc. in biology and biochemistry from the University of Ottawa, and has completed postbaccalaureate studies in ecology and wildlife management at the University of Alberta. He has worked in government and the private sector as an environmental scientist for more than twenty years, specializing in the investigation, evaluation, and mitigation of the effects of resource development on marine, wetland, and terrestrial ecosystems, organisms, and habitats. He has taught courses and lectured on subjects related to ecology and impact assessment at various technical schools and universities. He has testified as an expert witness in quasi-judicial hearings on the quality, completeness, and validity of studies filed in support of development projects.

Ann L. Riley has twenty years of experience in government. Her activities have included community organizing in the federal Office of Economic Opportunity, field work for the U.S. Geological Survey, land use planning for county governments in the Midwest, and water conservation, integrated pest management, river restoration, and floodplain management for the California Department of Water Resources. She is currently executive director of the Golden State Wildlife Federation. She has a Ph.D. from the University of California, Berkeley, specializing in floodplain management and stream restoration. She is completing a book on stream restoration to be published in 1993.

Rowan A. Rowntree since 1979 has been the research project leader for the study of forests and urbanization as part of the U.S. Forest Service's Northeastern Forest Experiment Station. Prior to that he was an associate professor of geography at Syracuse University and taught in the field of natural resources and environmental science. He was born and educated in California, with advanced degrees in both resource science and biogeography from the University of California at Berkeley. He is the author of over sixty research articles and presently supervises a national research program with staff and cooperating scientists at ten universities. Rowntree was a fellow with Resources for the Future, an advisor to the President's Commission on the National Parks and a visiting scholar at the University of California at Berkeley.

James A. Schmid has been an environmental consultant specializing in wetlands and environmental impact assessment for the last twenty years. He is an honor graduate of Columbia College in geography. He focused on vegetation patterns and landscape change in his graduate work at the University of Chicago Department of Geography. Schmid's master's research led him to the Edwards Plateau of southcentral Texas. His doctoral research on urban vegetation with special attention to metropolitan Chicago is still in print. During the early 1970s, Dr. Schmid taught on the faculty of biological sciences at Barnard College and Columbia University. His two-volume checklist of New Jersey plants was published in 1990.

Herbert W. Schroeder is an environmental psychologist working with the USDA Forest Service, North Central Forest Experiment Station, recreation research project in Chicago. He received his doctorate from the University of Arizona in 1980, and since then he has been doing research on people's perceptions, preferences, experiences, and values in forest environments.

Richard L. Whitman received his Ph.D. from Texas A&M University in wildlife and fisheries sciences. He was an associate professor of biology at Indiana University Northwest from 1979 to 1989. Since then he has been the chief scientist at the Research Division, Indiana Dunes National Lakeshore. His areas of interest include limnology, toxicology, and invertebrate ecology.